T0205469

Zero Index Metamaterials

Nishant Shankhwar · Ravindra Kumar Sinha

Zero Index Metamaterials

Trends and Applications

 Springer

Nishant Shankhwar
Department of Applied Physics
Delhi Technological University
Delhi, India

Ravindra Kumar Sinha
Department of Applied Physics
Delhi Technological University
Delhi, India

ISBN 978-981-16-0191-0   ISBN 978-981-16-0189-7   (eBook)
https://doi.org/10.1007/978-981-16-0189-7

This Springer imprint is published by the registered company Springer Nature Singapore Pte Ltd.
The registered company address is: 152 Beach Road, #21-01/04 Gateway East, Singapore 189721,
Singapore

Preface

This book is dedicated to an emerging regime of *zero refractive index photonics*, which involves metamaterials that exhibit effectively zero refractive index. Metamaterials are artificial structures whose optical properties can be tailored at will. With metamaterials, intriguing and spellbinding phenomena like negative refraction and electromagnetic cloaking could be realized, which otherwise seem unnatural or straight out of science fiction.

The metamaterial as perceived, analyzed, and presented in this book are governed by the principles of electrodynamics. Chapter 1 of the book presents a few basic concepts of electromagnetic theory which are prerequisite for diving into the realm of metamaterials, especially the zero refractive index variants. These include the evolution of electrodynamics, the physical interpretation of Maxwell's equations, the origin of refractive index and its manipulation using metamaterials, the reciprocal lattice and Brillouin zones, and a detailed account of the photonic crystals and their role as zero-index metamaterials. After getting basic insight into the needed concepts of electromagnetic theory, it becomes easy for the reader to grasp an advanced concept of zero-index metamaterial which is immediately introduced in Chap. 2 and continues in full rigor through out the book. Chapter 2 lays down the fundamental principle of zero refractive index property and the methods by which it can be achieved and verified. It discusses the popular zero-index metamaterials with their working principle and their influence on established phenomena like Snell's law and standing wave in a very abnormal yet absolutely physical manner. Afterward, Chap. 3 contains the most famous and impressive application of zero-index metamaterials, including electromagnetic tunneling, electromagnetic cloaking, wavefront engineering, and directive radiation. Zero-index metamaterials are also seen as means of boosting nonlinear properties and are believed to have strong prospects for being useful in nonlinear optics. Chapter 4 deals with these possibilities and presents various nonlinear phenomena which zero-index metamaterials have eased or enhanced, or have to potential to enhance. Finally, Chap. 5 discusses the recent advancements happened in this arena and the future prospects this newly emerging field holds, respectively.

We have decided to write this book for the benefit of freshmen researchers, who are interested in this appealing and largely unexplored field but find it difficult to acquire sufficient and suitable literature to learn from. This book throws light on all the aspects of zero refractive index phenomenon ranging from its advent to current achievements, applications, and future possibilities. We have included elaborated working details and their rigorous numerical verification for the sake of profound understanding of an ardent reader. In a nutshell, this text accumulates almost everything that is available on zero-index metamaterials, in one place, and will hopefully prove useful for a professionally interested and motivated reader.

Delhi, India   Nishant Shankhwar
                                                                                   Ravindra Kumar Sinha

Acknowledgments

The authors acknowledge the initiatives and support from the establishment of TIFAC-Centre of Relevance and Excellence in Fiber Optics and Optical Communication at the Department of Applied Physics, Delhi Technological University (Formerly Delhi College of Engineering), Delhi, through the Mission REACH program of Technology Vision-2020 of the Government of India. One of the authors (NS) wishes to thank his family for always having faith in him and motivating him at the times of *"crests and troughs"* during this project. The other author (RKS) wishes to acknowledge the support by his family members, his teachers, and doctoral students for their personal and professional support in preparing this book. Both the authors would like to thank Professor Yogesh Singh, Vice Chancellor, DTU, and all the faculty members of Applied Physics Department of Delhi Technological University, Delhi, for providing conducive environment for carrying out the research work in the emerging area of zero-index metamaterials.

Nishant Shankhwar
Ravindra Kumar Sinha

Contents

Abbreviations

AZO	Aluminum zinc oxide
BZ	Brillouin zone
CMOS	Complimentary metal-oxide semiconductor
DFG	Difference frequency generation
EMNZ	Epsilon-and-mu-near-zero
ENZ	Epsilon-near-zero
FDM	Finite difference method
FDTD	Finite difference time domain
FEM	Finite element method
FF	Fundamental frequency
GH	Goos–Hanchen
IBZ	Irreducible Brillouin zone
IPM	Imperfect phase matching
ITO	Indium tin oxide
MM	Metamaterial
MNZ	Mu-near-zero
MPB	MIT photonic bands
NLO	Nonlinear optics
NZPI	Negative-zero-positive-index
PBG	Photonic bandgap
PEC	Perfect electric conductor
PhC	Photonic crystals
PM	Phase matching
PMC	Perfect magnetic conductor
PML	Perfectly matched layer
PMMA	Polymethyl methacrylate
PPM	Perfect phase matching
PWEM	Plane wave expansion method
QPM	Quasi-phase matching
SGF	Sum frequency generation
SHG	Second harmonic generation
TCO	Transparent conducting oxide

TE	Transverse electric
TM	Transverse magnetic
ZIM	Zero-index metamaterial

Chapter 1
Electromagnetics for Zero-Index Metamaterials

1.1 History of Electrodynamics

The nineteenth century witnessed tremendous progress in electromagnetism, prior to which electricity and magnetism were not perceived as the two sides of the same coin, as they are today. It all started in 1820, when Hans Christian Øersted demonstrated for the first time that an electric current flowing in a wire generated a circulating magnetic field around it (Fig. 1.1a) [1]. In the following year, André-Marie Ampère took things further by showing that two parallel current-carrying wires exerted attractive or repulsive force on each other depending upon the direction of current in one w.r.t. the other (Fig. 1.1b) [1–3]. Besides, he also presented a theoretical explanation of the phenomenon and established a mathematical relation between the magnitude of the current flowing and the intensity of the magnetic field generated. As the news about Øested's experiment reached Michael Faraday in the Royal Society, he immediately conceived the idea of a device that could harness the magnetic energy and furnish some mechanical work. This led to the invention of the primordial electric motor by him in the year 1821 (Fig. 1.1c). Moreover, Faraday envisioned the possibility of realizing the reverse phenomenon. He ardently believed that symmetry was nature's innate property, and hence if electricity could produce magnetism, the vice versa must be true as well. A series of painstaking experiments during the next 10 years led to the discovery of electromagnetic induction in 1931 [4, 5]. By this time, the coalition between electricity and magnetism had been established. Another significant milestone in this arena was set by James Clerk Maxwell, who established light as a manifestation of the principles of electrodynamics. He argued that a varying electric field generated a varying magnetic field, which in turn generated another varying electric field and the phenomenon perpetuated [6]. He brilliantly condensed entire electromagnetic theory into four simple equations, now referred to as Maxwell's equation. These equations are the most generalized mathematical representations of Gauss's law for electricity and magnetism, Faraday's law, and Ampere's circuital

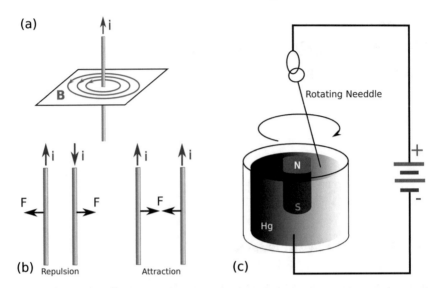

Fig. 1.1 a Circular magnetic field lines around a current-carrying conductor; **b** force exerted by two parallel current-carrying conductors on each other depending upon the direction of current flow in them; **c** the first electric motor invented by Faraday

law. He also introduced a significant correction factor in Ampere's law called the *displacement current*, which facilitated the comprehension of electromagnetic waves.

Having a physical and mathematical understanding of electromagnetic waves made it easier to interpret their interaction with matter. The response of a material to an electromagnetic wave is linked to the redistribution of charges inside it, upon being subject to the electric and magnetic fields of the wave. This knowledge about light–matter interactions led to the development of numerous devices working in different parts of the spectrum, during the entire twentieth century. The transition period from the twentieth to the twenty-first century, i.e., late 1990s and early 2000s, marked the beginning of a new era of electromagnetics by the advent of a special class of man-made materials called *metamaterials*, which have several abnormal and unnatural properties like negative refraction, electromagnetic cloaking, perfect lensing, etc. This chapter encompasses all those fundamental principles of electromagnetics which are needed as a prerequisite for efficient comprehension of advanced concepts spread throughout the book. By reading this chapter, even a college freshman can build the ability to grasp the physics of metamaterials to his complete intellectual satisfaction. The first step on this route should be Maxwell's equation, but to understand them, one must first be acquainted with the physical meaning of the terms—gradient, divergence, and curl, which has been explained in Section [2, 3, 7].

1.2 Physical Meaning of Gradient, Divergence, and Curl

The aim of this section is not to teach the formulae and calculations of gradient, divergence, and curl for various electromagnetic systems, as they are already present in several good quality texts like Feynman's Lectures Vol. 2 [2], Griffiths [3], Sadiku [7], and Spiegel [8]. Here we intend to present only the physical meanings of these popular terms of vector calculus, in order to assist the reader's imagination to visualize the behavior of electromagnetic waves in different types of systems that shall be encountered throughout this book.

1.2.1 Gradient

Whenever one comes across the term gradient, he should immediately think of a quantity, such as height, increasing or decreasing in a "particular direction." English meaning of the term gradient is *slope*, as in the case of a hill. For example, as shown in Fig. 1.2a, when an object rolls down a hill its height (h) from the ground level reduces, whereas during uphill motion (Fig. 1.2b) it increases. In both the cases, there is a motion and distance traveled along x-direction too. Here, "height" is the quantity which is varying with respect to x. The variation in case (a) (downhill motion) is negative while in case (b) is positive, and hence gradient (i.e., *slope*) $\partial h/\partial x$ is negative in the former while positive in the latter case. By closer inspection, an additional piece of information is obtained from Fig. 1.2. One should notice that in the shown physical system, the increase or decrease in the height h is taking place along a particular direction, "x" in this case. Although height is a scalar quantity, gradient is a vector! Other scalar physical quantities dependent on height, such as potential energy ($V = mgh$), follow the same trend and exhibit similar behavior of negative gradient in case (a) and positive gradient in case (b). In general, the gradient of any scalar physical quantity is a vector, since it has a direction and can be shown to abide by the laws of vector addition.

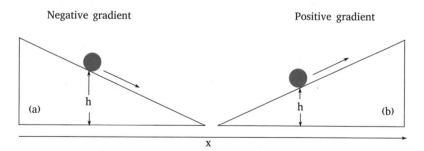

Fig. 1.2 Physical meaning of gradient

1.2.2 Divergence

According to dictionary, diverge means *to separate* and converge means *to come together*. In electromagnetics, the divergence has special significance in relation to the distribution and orientation of the electric field. The electric field, in simple terms, is the influence of a charge felt in its surroundings. In close proximity to the charge, the influence is strong and becomes gradually weaker as one moves away from it. The term divergence describes the nature of the field around the charge. To visualize the true nature of the electric field, let us think of an isolated fixed positive charge $+Q$ which influences a free small positive test charge $+q$ present in its vicinity. *How does $+Q$ affect $+q$?* See Fig. 1.3a for the answer. Due to repulsive force between two positive charges the test charge $+q$ will be pushed radially away from $+Q$, and as it travels farther, the influence of $+Q$ on it weakens. In light of this observation, it became a convention to graphically represent the electric field of a positive charge by means of arrows pointing radially outward from it, as shown in Fig. 1.3a. It can also be seen as the trajectory of $+q$, under the influence of $+Q$.

Similarly, the electric field of a fixed negative charge $-Q$ is represented by arrows pointing radially inward, as shown in Fig. 1.3b, since $-Q$ tends to attract $+q$ toward it. It should be noted that in both the cases, the field lines are denser near the fixed charge and tend to rarefy away from it, which symbolizes the reduction in the field's magnitude as one moves away from the charge. In the third case of an electric dipole (Fig. 1.3c), where two equal and opposite charges are separated from each other by a certain distance, the electric field lines originate from the positive charge and sink into the negative charge. The number of field lines shown for both the charges is the same because they have been assumed to be of equal magnitude.

Now, if one determines the divergence of the electric field, he always obtains a positive value for the "field due to a positive charge," since its electric field lines are literally diverging (Fig. 1.3a). In other words, the field due to a positive charge has a positive divergence. On the other side, the field lines around a negative charge (Fig. 1.3b) are always converging into it, hence, it is considered to have negative divergence. On this basis, it can be stated that an electric dipole should have zero divergence since all the field lines diverging from the positive end are eventually converging into the negative end. Hence, no net field lines are actually emerging out

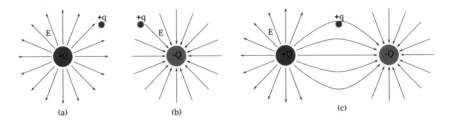

(a)   (b)   (c)

Fig. 1.3 Physical meaning of divergence

or sinking into the system. *One may wonder, why divergence (and not convergence) has been embraced as a property of the field?* Answer to this is the fact that since the advent of electrodynamics positive charge has conventionally been pivotal in defining various quantities and parameters. It was not known back then that electron (a negatively charged particle) was the fundamental unit of charge, hence the convention.

1.2.3   Curl

Curl, as the name suggests, is the measure of the twist or swirl of a vector field. To understand the physical meaning of curl, let us think of a paddled wheel submerged in three different types of streams of water. Figure 1.4 shows the bird's-eye view of this thought experiment. In the first type, shown in Fig. 1.4a, water has uniform velocity across the stream, and hence both the paddles (1 and 2) will experience equal thrust by water. As a result, the paddled wheel will translate with the flow of water but will not rotate, since it experiences a net force, but no net torque. In this case, the velocity vector field has no swirl at all, and hence its curl is zero.

The second case shown in Fig. 1.4b is slightly different, where the velocity of water goes on decreasing from one bank to the other. In this situation, the force exerted on both the paddles is in the same direction, but of different magnitudes. Force exerted on the paddle 1 is greater than that on paddle 2, on account of the greater velocity of water near the former than the latter. As a result, this time the paddled wheel will experience a net torque alongside a net force, and hence will exhibit both translational and rotational motions. In this case, the velocity vector field has a finite curl, albeit of a low magnitude.

The third case is of a more complex nature. Here, the water flow becomes weaker from a bank to the middle of the stream, the velocity becomes zero at the exact center, the direction of the flow reverses beyond that, and the velocity goes on increasing till the other bank. In other words, the magnitude of the velocity of water increases symmetrically on both sides from the center, but along opposite directions. In this

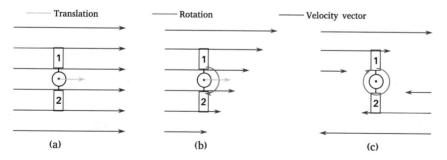

Fig. 1.4 Physical meaning of curl

scenario, the force on the two paddles will be the same, but along opposite directions. Consequently, the wheel will experience no net force but only a net torque, and will exhibit rotation but no translatory motion. It should be noted that in this case, the curl of the vector field will be relatively greater than that in case (b), since the direction of water flow is opposite in the two halves of the stream. Now, if the velocity is replaced by any other vector quantity such as electric field or magnetic field, the meaning of curl remains unchanged. In light of the above explanations, we believe that the reader will develop a sufficient insight into the physical significance of gradient, divergence, and curl, which will be invoked in context to Maxwell's equations and the electromagnetic waves.

1.3 Maxwell's Equation

Maxwell's equations are a brilliant example of unification and generalization. The entire electrodynamics is packed into a simple set of these four mathematical equations. Below are the different variants of Maxwell's equations subject to certain conditions [2, 3, 7, 9].

Differential form

$$\nabla \cdot \mathbf{E} = \rho/\epsilon \tag{1.1}$$

$$\nabla \cdot \mathbf{B} = 0 \tag{1.2}$$

$$\nabla \times \mathbf{E} = -\frac{\partial \mathbf{B}}{\partial t} \tag{1.3}$$

$$\nabla \times \mathbf{B} = \mu \mathbf{J}_s + \mu\epsilon \frac{\partial \mathbf{E}}{\partial t} \tag{1.4}$$

Integral form

$$\oiint \mathbf{D} \cdot \mathbf{ds} = \iiint \rho dv \tag{1.5}$$

$$\oiint \mathbf{B} \cdot \mathbf{ds} = 0 \tag{1.6}$$

$$\oint \mathbf{E} \cdot \mathbf{dl} = -\frac{\partial}{\partial t} \iint \mathbf{B} \cdot \mathbf{ds} \tag{1.7}$$

$$\oint \mathbf{B} \cdot \mathbf{dl} = \mu \iint \mathbf{J}_s \cdot \mathbf{ds} + \mu\epsilon \frac{\partial}{\partial t} \iint \mathbf{E} \cdot \mathbf{ds} \tag{1.8}$$

Special cases

Static field assumption: On assuming the fields to be invariant with time, i.e., $\partial/\partial t = 0$, Eqs. 1.1–1.4 reduce down to

$$\nabla \cdot \mathbf{E} = \rho/\epsilon \tag{1.9}$$

$$\nabla \cdot \mathbf{B} = 0 \tag{1.10}$$

$$\nabla \times \mathbf{E} = 0 \tag{1.11}$$

$$\nabla \times \mathbf{B} = \mu \mathbf{J}_s \tag{1.12}$$

Time-harmonic field assumption: On assuming the time-harmonic variation, i.e., field varying as $e^{-i\omega t}$ with respect to time, $\partial/\partial t$ can be replaced by $-i\omega$ and Eqs. 1.1–1.4 reduce down to

$$\nabla \cdot \mathbf{E} = \rho/\epsilon \tag{1.13}$$

$$\nabla \cdot \mathbf{B} = 0 \tag{1.14}$$

$$\nabla \times \mathbf{E} = i\omega \mathbf{B} \tag{1.15}$$

$$\nabla \times \mathbf{B} = \mu \mathbf{J}_s - \mu \epsilon i \omega \mathbf{E} \tag{1.16}$$

where

$\mathbf{E}$ = electric field vector,
$\mathbf{B}$ = magnetic field vector,
$\mathbf{J}_s$ = current density,
dv = volume element,
$\mathbf{ds}$ = area element,
$\mathbf{dl}$ = length element,
μ = permeability,
ϵ = permittivity,
ρ = charge density,
ω = frequency,
$i = \sqrt{-1}$.

Interpretation of Maxwell's equations: The first equation is Gauss's law which states that the divergence of an electric field is proportional to the charge density. The second equation is the magnetic equivalent of Gauss's law which invalidates the existence of magnetic monopoles. The absence of magnetic monopoles is attributed to the fact that at the very fundamental level of magnetism exists a dipole. An electron revolving around the nucleus acts as a current-carrying loop which has the magnetic field distribution of a dipole. The third equation is Faraday's law which states that a time-varying magnetic field generates a space-varying electric field and became the basis of electromagnetic induction. The fourth equation is Ampere–Maxwell law which states that a magnetic field can be generated by a steady-state current as well as by a time-varying electric field. For a more detailed account of Maxwell's equations, the reader is advised to read Griffiths [3]. Moreover, Maxwell's fourth equation also indicates the electromagnetic nature of light. It is now understood that electromagnetic wave propagation is possible because of an interplay of Eqs. 1.3 and 1.4, where a varying electric field produces a varying magnetic field, which in turn produces a varying electric field and the phenomenon perpetuates. The electric

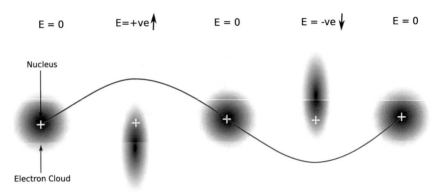

Fig. 1.5 The electron cloud shifts according to the magnitude and direction of the electric field of the electromagnetic wave. Illustrated here is the displacement of electron cloud through one complete cycle of oscillation of the electric field

permittivity and the magnetic permeability offered by different materials depend on light–matter interaction. The next section throws light on this aspect and explains the cause of refractive index.

1.4   Origin of Refractive Index

The refractive index offered by a material depends on how its atoms interact with light, which further depends on how strongly the electrons are bound to the nucleus. Different materials have different values of refractive index because the strength by which the electrons are bound to the nucleus is different in each one of them. Moreover, for a particular material, the refractive index is a strong function of the wavelength of the incident light, especially in the optical region [10–13]. A rigorous analysis of the dependence of refractive index on the material-specific intrinsic parameters and the wavelength of the incident light is given below.

Let a given medium be exposed to an x-polarized light. The electric field associated with the electromagnetic wave is

$$\mathbf{E} = \hat{\mathbf{x}}E_0 cos(kz - \omega t) \tag{1.17}$$

The electric field causes the electron cloud to get displaced opposite to the direction of the electric field vector (as shown in Fig. 1.5), resulting in polarization of the atom. The displacement (say $\mathbf{x}$) of the center of the electron cloud with respect to the nucleus is given by the solution of the differential equation

$$m\frac{d^2\mathbf{x}}{dt^2} + k_0\mathbf{x} = -q\mathbf{E} \tag{1.18}$$

or

$$\frac{d^2\mathbf{x}}{dt^2} + \omega_0^2\mathbf{x} = -\frac{q\mathbf{E}}{m} \tag{1.19}$$

where $-q$ and m are the charge and the mass of an electron, k_0 is the restoring force constant, and $\omega_0 = \sqrt{k_0/m}$ is the resonant frequency. On solving Eq. 1.18, we get

$$\mathbf{x} = -\frac{q\mathbf{E}}{m(\omega_0^2 - \omega^2)} \tag{1.20}$$

If the number of dispersion electrons per unit volume is N, then the polarization $\mathbf{P}$ induced in the medium by electric field $\mathbf{E}$ is given by

$$\mathbf{P} = -Nq\mathbf{x} \tag{1.21}$$

$$= \frac{Nq^2}{m(\omega_0^2 - \omega^2)}\mathbf{E} \tag{1.22}$$

$$= \epsilon_0\chi\mathbf{E} \tag{1.23}$$

where

$$\chi = \frac{Nq^2}{m\epsilon_0(\omega_0^2 - \omega^2)} \tag{1.24}$$

is the electric susceptibility and ϵ_0 is the permittivity of free space. Hence, the permittivity of the medium is given by

$$\epsilon = \epsilon_0 + \epsilon_0\chi \tag{1.25}$$

$$= \epsilon_0(1 + \chi) = \epsilon_0\epsilon_r \tag{1.26}$$

where ϵ_r is the relative permittivity of the medium given by

$$\epsilon_r = 1 + \chi = 1 + \frac{Nq^2}{m\epsilon_0(\omega_0^2 - \omega^2)} \tag{1.27}$$

It is known that the relative permittivity is square of the refractive index, and hence Eq. 1.27 can be written as

$$n^2 = 1 + \chi = 1 + \frac{Nq^2}{m\epsilon_0\omega_0^2}\left(1 - \frac{\omega^2}{\omega_0^2}\right)^{-1} \tag{1.28}$$

$$\approx 1 + \frac{Nq^2}{m\epsilon_0\omega_0^2}\left(1 + \frac{\omega^2}{\omega_0^2}\right) \tag{1.29}$$

$$\approx 1 + \frac{Nq^2}{m\epsilon_0\omega_0^2} + \frac{Nq^2}{m\epsilon_0\omega_0^4}\frac{4\pi^2c^2}{\lambda_0^2} \tag{1.30}$$

$$= A + \frac{B}{\lambda_0^2} \tag{1.31}$$

which is the well-acknowledged *Cauchy's formula*, expressing refractive index as a function of wavelength [14].

In the case of metals, the electrons are free and there is no restoring force acting on them. Hence, $\omega_0 = 0$, and Eq. 1.27 is reduced to

$$n^2 = 1 - \frac{Nq^2}{m\epsilon_0\omega^2} \tag{1.32}$$

$$= 1 - \frac{\omega_p^2}{\omega^2} \tag{1.33}$$

where

$$\omega_p = \sqrt{\frac{Nq^2}{m\epsilon_0}} \tag{1.34}$$

is the plasma frequency of the metal. Below plasma frequency, i.e., for $\omega < \omega_p$, the refractive index is imaginary which accounts for the ohmic loss associated with metals. Above plasma frequency, i.e., for $\omega > \omega_p$, the refractive index is real and metals behave like dielectrics [15].

Please note that in the above analysis the damping force has not been considered. Considering the damping factor is necessary because it does become significant in a certain region of the spectrum. For example, silicon is a lossy dielectric in the visible region but is practically lossless in the infrared (e.g., 1550 nm) and beyond. Hence, in a more realistic perception, Eq. 1.18 should be written as

$$m\frac{d^2\mathbf{x}}{dt^2} + \Gamma\frac{d\mathbf{x}}{dt} + k_0\mathbf{x} = -q\mathbf{E} \tag{1.35}$$

and Eq. 1.19 becomes

$$\frac{d^2\mathbf{x}}{dt^2} + 2K\frac{d\mathbf{x}}{dt} + \omega_0^2\mathbf{x} = -\frac{q\mathbf{E}}{m} \tag{1.36}$$

Solving Eq. 1.36, the accurate expressions for polarization, susceptibility, and relative permittivity, inclusive of the damping factor $K = \Gamma/2m$, are obtained as

$$\mathbf{P} = \frac{Nq^2}{m(\omega_0^2 - \omega^2 - 2iK\omega)}\mathbf{E}, \tag{1.37}$$

$$\chi = \frac{Nq^2}{m\epsilon_0(\omega_0^2 - \omega^2 - 2iK\omega)} \tag{1.38}$$

and

$$\epsilon_r = 1 + \frac{Nq^2}{m\epsilon_0(\omega_0^2 - \omega^2 - 2iK\omega)} \tag{1.39}$$

It can be noticed that the relative permittivity is complex in nature, which implies that the refractive index is complex too. Let the complex refractive index be represented as

$$n = \eta + i\kappa \tag{1.40}$$

Then,

$$n^2 = \epsilon_r = (\eta + i\kappa)^2 = \eta^2 - \kappa^2 + 2i\eta\kappa \tag{1.41}$$

On comparing Eqs. 1.39 and 1.41, we get

$$(\eta + i\kappa)^2 = 1 + \frac{Nq^2}{m\epsilon_0(\omega_0^2 - \omega^2 - 2iK\omega)} \tag{1.42}$$

$$= 1 + \frac{Nq^2(\omega_0^2 - \omega^2 + 2iK\omega)}{m\epsilon_0(\omega_0^2 - \omega^2 - 2iK\omega)(\omega_0^2 - \omega^2 + 2iK\omega)} \tag{1.43}$$

or

$$\eta^2 - \kappa^2 = 1 + \frac{Nq^2(\omega_0^2 - \omega^2)}{m\epsilon_0[(\omega_0^2 - \omega^2)^2 + 4K^2\omega^2]} \tag{1.44}$$

and

$$2\eta\kappa = \frac{Nq^2(2K\omega)}{m\epsilon_0[(\omega_0^2 - \omega^2)^2 + 4K^2\omega^2]} \tag{1.45}$$

Figure 1.6 shows the qualitative variation of the real and imaginary parts of the relative permittivity with respect to frequency for both the dielectrics and the metals.

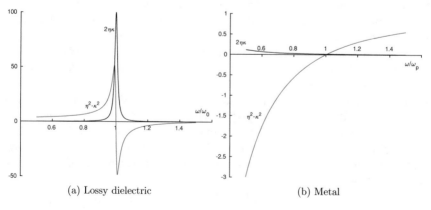

(a) Lossy dielectric   (b) Metal

Fig. 1.6 Qualitative variation of the real and imaginary parts of the relative permittivity with respect to frequency for **a** dielectrics and **b** metals (assuming $\omega_0 = 0$)

Since metals have free electrons, resonant frequency ω_0 is zero. The x-axis of the graph has been normalized w.r.t. the resonant frequency ω_0 in Fig. 1.6a and w.r.t. the plasma frequency ω_p in Fig. 1.6b. Below the plasma frequency, $\eta^2 - \kappa^2$ is negative and $2\eta\kappa$ is significantly high, which accounts for the highly reflective and absorptive nature of metals. Above the plasma frequency, $\eta^2 - \kappa^2$ becomes positive, thus metals begin to acquire dielectric-like properties of wave propagation. Additionally, the value of $2\eta\kappa$ also descends to extremely low values which in turn reduces absorption losses. Hence, for $\omega > \omega_p$ metals become transparent to electromagnetic waves.

1.5   Structure-Dependent Refractive Index

In the previous section, the origin of refractive index of natural materials was discussed, which was dependent on how strongly the electrons are held by their nucleus, and hence how strongly they could respond to the electric field of the incident light. However, there is a special class of artificial materials called the *metamaterials*, for which the optical parameters like relative permittivity, relative permeability and refractive index depend not only on their constituent materials but also on their structure [16–35]. Their peculiar structure imparts to them the exotic properties, which are drastically different from those of their constituent materials. Presented below is an example of one such metamaterial structure, which comprises the alternating layers of a metal and a dielectric material, one of the easiest ways to achieve the desired permittivity at the wavelength of interest.

Metals intrinsically exhibit negative dielectric constant below their plasma frequency due to the presence of free electrons in them. This property proves beneficial in realization of negative refractive index where negative permittivity ($\epsilon < 0$) and negative permeability ($\mu < 0$) are required at the same wavelength. Interleaving of dielectric layers in between metallic layers dilutes the metal in a way and allows the liberty to reduce the permittivity to the desired magnitude [15]. The value of effective permittivity thus obtained depends on the filling fraction of the metal and the dielectric in a unit cell as well as on the polarization of the incident field. For the structure shown in Fig. 1.7, let us suppose that ϵ_m is the permittivity of the metal, ϵ_d is the permittivity of the dielectric, h_m and h_d are their respective thicknesses, and in consequence, $f_m = h_m/(h_m + h_d)$ and $f_d = h_d/(h_m + h_d) = 1 - f_m$ become the filling fractions of the metal and the dielectric, respectively. The incident electric field can either be polarized parallel or perpendicular to the interface [23]. In the case of perpendicular polarization, the electric field is discontinuous across the boundary, but the displacement vector is continuous, i.e., $D_m = D_d = D_\perp$. On account of discontinuity, the effective electric field needs to be taken as a weighted mean and is thus given by

$$E_\perp = f_m E_m + f_d E_d \tag{1.46}$$

Hence, the effective permittivity $\epsilon_\perp$ is given by

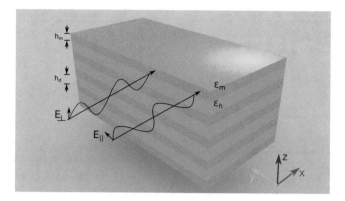

Fig. 1.7 Schematic illustration of metal-dielectric-layer-based electric metamaterial, illuminated by parallel and perpendicular polarization

$$\frac{1}{\epsilon_\perp} = \frac{E_\perp}{D_\perp} \tag{1.47}$$

$$= \frac{f_m E_m}{D_\perp} + \frac{f_d E_d}{D_\perp} \tag{1.48}$$

$$= \frac{f_m}{\epsilon_m} + \frac{f_d}{\epsilon_d} \tag{1.49}$$

On the contrary, for parallel polarization, the electric field is continuous across the interface, i.e., $E_m = E_d = E_\parallel$, whereas the displacement vector is discontinuous. Hence, in this case, the effective displacement vector needs to be taken as a weighted mean, given by

$$D_\parallel = f_m D_m + f_d D_d \tag{1.50}$$

Therefore, the effective permittivity is

$$\epsilon_\parallel = \frac{D_\parallel}{E_\parallel} \tag{1.51}$$

$$= \frac{f_m D_m}{E_\parallel} + \frac{f_d D_d}{E_\parallel} \tag{1.52}$$

$$= f_m \epsilon_m + f_d \epsilon_d \tag{1.53}$$

The effective permittivity as a function of wavelength, for the perpendicular and parallel polarization, has been shown in Figs. 1.8 and 1.9. Looking at these figures, it is clear that the effective permittivity, hence the effective refractive index, can be controlled at will by controlling the structural parameters.

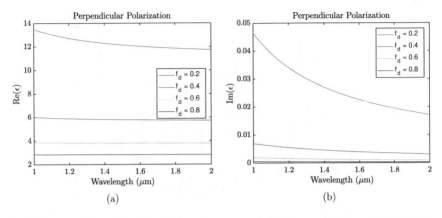

Fig. 1.8 Dependence of (a) real and (b) imaginary parts of effective permittivity on dielectric filling fraction for perpendicular polarization

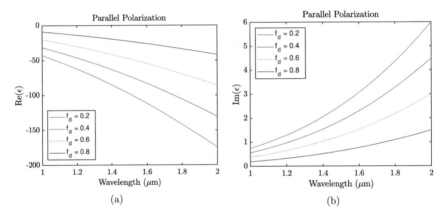

Fig. 1.9 Dependence of **a** real and **b** imaginary parts of effective permittivity on dielectric filling fraction for parallel polarization

1.6 Reciprocal Lattice and Brillouin Zone

Every crystal has two lattices associated with it—a real lattice and a reciprocal lattice. The primitive vectors of the reciprocal lattice are reciprocal of the real lattice's primitive vectors. The unit cell of the reciprocal lattice is called a Brillouin zone (BZ). If the basis of a real lattice is made up of position vectors, the reciprocal lattice is constituted of wave vectors. A reciprocal lattice is basically the diffraction pattern or Fourier transform of the real lattice [36, 37].

A unit cell of an arbitrary lattice and its Brillouin zone have been shown in Fig. 1.10. From the basic knowledge of vector algebra we know that the area of the face including a_1 and a_2 is given by $\mathbf{A} = \mathbf{a_1} \times \mathbf{a_2}$ and the volume of the unit cell is given by $V = (\mathbf{a_1} \times \mathbf{a_2}) \cdot \mathbf{a_3}$. It is understandable that the vector $\mathbf{a_3}$ is written in

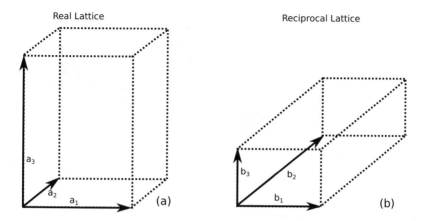

Fig. 1.10 Unit cells of **a** real lattice and **b** reciprocal lattice

terms of V and A as, $\mathbf{a_3} = V/A$, hence the reciprocal vector $\mathbf{b_3} = A/V = (\mathbf{a_1} \times \mathbf{a_2})/[(\mathbf{a_1} \times \mathbf{a_2}) \cdot \mathbf{a_3}]$. In this way, all the three primitive reciprocal vectors can be defined as [8, 36, 37]

$$\mathbf{b_1} = 2\pi \frac{\mathbf{a_2} \times \mathbf{a_3}}{\mathbf{a_1} \cdot \mathbf{a_2} \times \mathbf{a_3}} \qquad (1.54)$$

$$\mathbf{b_2} = 2\pi \frac{\mathbf{a_3} \times \mathbf{a_1}}{\mathbf{a_1} \cdot \mathbf{a_2} \times \mathbf{a_3}}$$

$$\mathbf{b_3} = 2\pi \frac{\mathbf{a_1} \times \mathbf{a_2}}{\mathbf{a_1} \cdot \mathbf{a_2} \times \mathbf{a_3}}$$

It should be noted that the factor of 2π is included for the sake of convenience so that the reciprocal vector can be translated into the wave vector.

A crystal lattice is a periodic arrangement of atoms, which means a periodic variation of potential. This idea has been extended to optics, and certain artificial structures have been developed by achieving periodic distribution of refractive index, called the *photonic crystals* [37–39]. The most common type of photonic crystal is a square array of dielectric rods in air (as shown in Fig. 1.11a). It can be noticed from Fig. 1.11b that a square real lattice has a square reciprocal lattice, with a square Brillouin zone (Fig. 1.11b and c). In the Brillouin zone, both the components of wave vector (k_x and k_y) vary from $-\pi/a$ to π/a. Closer inspection tells that the Brillouin zone is not the most fundamental unit, since there is a smaller and more fundamental unit of the reciprocal lattice, shown as the wedge-shaped shaded area in Fig. 1.11c. This shaded area is referred to as an *irreducible Brillouin zone* (IBZ) and the rest of the BZ can be obtained from it, by applying mirror and rotation symmetry. For the intellectual satisfaction of a more ardent reader, Kittle [36] presents an exhaustive explanation of the reciprocal lattice and Brillouin zone for a variety of two- and three-dimensional crystal structures.

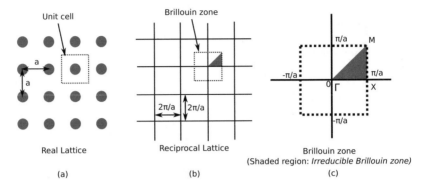

Unit cell   Brillouin zone

Real Lattice   Reciprocal Lattice   Brillouin zone
(Shaded region: *Irreducible Brillouin zone*)

(a)   (b)   (c)

Fig. 1.11 **a** A square lattice of dielectric rods, **b** its square reciprocal lattice, and **c** Brillouin zone and the irreducible Brillouin zone

1.7 Photonic Crystals

The purpose of discussing the concept of reciprocal lattice and the Brillouin zone in the previous section was to ensure easy comprehension of *photonic crystals*. Photonic crystals (PhC) are artificial materials that have a periodic variation of refractive index in one, two, or three dimensions. They influence the incoming photons in a similar manner as the crystals do to the incoming electrons; however, the difference is that the former has a periodic variation of potential while the latter has a periodic distribution of the refractive index. Similar to the electronic band structure for crystals, photonic band structures can be calculated for photonic crystals too [37, 40]. Shown below are the three types of photonic crystals based on the dimensions of periodicity. Figure 1.12a shows a one-dimensional photonic crystal, also called *Bragg mirror* [37], in which the refractive index varies periodically only in one dimension and is invariant in the remaining two dimensions. Similarly, the variation of refractive index in two and three dimensions yields two-dimensional and three-dimensional photonic crystals shown in Fig. 1.12b and c, respectively.

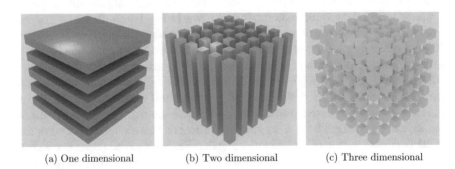

(a) One dimensional   (b) Two dimensional   (c) Three dimensional

Fig. 1.12 The three types of photonic crystals based on dimensions

1.7.1 *Maxwell's Equations Inside Photonic Crystals*

Photonic crystals are all-dielectric structures and do not sustain localized charges and steady-state currents, and hence the time-harmonic ($e^{-i\omega t}$) version of Maxwell's equations can be written as

$$\nabla \cdot \mathbf{E}(\mathbf{r}) = 0 \tag{1.55}$$

$$\nabla \cdot \mathbf{H}(\mathbf{r}) = 0 \tag{1.56}$$

$$\nabla \times \mathbf{E}(\mathbf{r}) = i\omega\mu\mathbf{H}(\mathbf{r}) \tag{1.57}$$

$$\nabla \times \mathbf{H}(\mathbf{r}) = -i\omega\epsilon_0\epsilon(\mathbf{r})\mathbf{E}(\mathbf{r}) \tag{1.58}$$

Using these equation, we obtain the following *master equation* [37]:

$$\nabla \times \left(\frac{1}{\epsilon(\mathbf{r})} \nabla \times \mathbf{H}(\mathbf{r}) \right) = \left(\frac{\omega}{c} \right)^2 \mathbf{H}(\mathbf{r}) \tag{1.59}$$

This equation is an eigenvalue equation which can be solved to obtain the eigenvalues ω and the corresponding eigenfunctions $\mathbf{H}(\mathbf{r})$. Electric field $\mathbf{E}(\mathbf{r})$ can then be determined from $\mathbf{H}(\mathbf{r})$ by

$$\mathbf{E}(\mathbf{r}) = \frac{i}{\omega\epsilon_0\epsilon(\mathbf{r})} \nabla \times \mathbf{H}(\mathbf{r}) \tag{1.60}$$

One must understand that the eigenfunction of the above master equation is not of a plane wave type but of a Bloch wave kind, i.e., a plane wave ($e^{i\mathbf{k}\cdot\mathbf{r}}$) times a periodic function ($u_\mathbf{k}(\mathbf{r})$) which inherits the periodicity of the photonic crystal. Hence,

$$\mathbf{H}(\mathbf{r}) = e^{i\mathbf{k}\cdot\mathbf{r}} u_\mathbf{k}(\mathbf{r}) \tag{1.61}$$

where $u_\mathbf{k}(\mathbf{r})$ satisfies $u_\mathbf{k}(\mathbf{r}) = u_\mathbf{k}(\mathbf{r} + \mathbf{T})$ if $\mathbf{T}$ is the periodicity of the photonic crystal. The master equation can be solved by numerical techniques such as *plane wave expansion method* (PWEM) [41, 42], finite difference method (FDM) [43, 44], finite element method (FEM) [45], etc. [9, 46] to obtain the photonic band diagram. The photonic bands are dispersion curves (ω versus k plots) for different eigenvalues. All the photonic band diagrams shown in this book have either been obtained by COMSOL Multiphysics [47] or by MIT's open-source software package MIT Photonic Bands (MPB) [48].

1.7.2 *Photonic Band Structure*

A photonic band structure is a photonic analog of the electronic band structure. As said above, it is obtained by solving the master equation. Below are the photonic bands

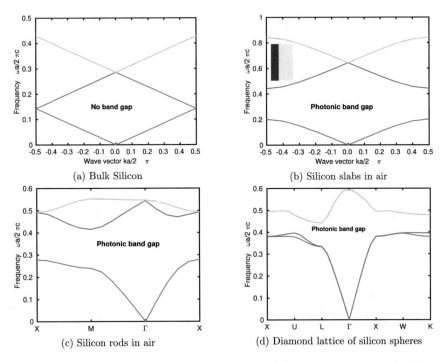

Fig. 1.13 Comparison of the band structure of three types of photonic crystal. Inset: Unitcell

of a variety of structures ranging from bulk material to three-dimensional diamond-lattice photonic crystal, computed using MPB. Similar to an electronic bandgap, a photonic band structure has a photonic bandgap region in which no propagating mode exists. Firstly, Fig. 1.13a shows the band structure for a homogeneous silicon medium, which is nothing but the light line for silicon folded back into the Brillouin zone, due to the periodic nature of the wave vector with a period of $2\pi/a$. This is a compact way of representing the dispersion relation based on the fact that all the possible modes are already present within the Brillouin zone. It can be seen that the bands are continuous without any discontinuity. Figure 1.13b shows the band diagram for a one-dimensional photonic crystal, which is made up of a *quarter-wave* stack of silicon and air regions. A quarter-wave stack means that the thickness of each layer is equal to one-fourth of the wavelength in that medium. One-dimensional photonic crystals of such design show maximum bandgap. A broad bandgap region can be observed in Fig. 1.13b. Similarly, Fig. 1.13c and d shows the band diagrams for square lattice of silicon rods in air (a 2D photonic crystal) and a diamond (i.e., face-centered cubic) lattice of silicon spheres (a 3D photonic crystal). A well-pronounced bandgap can be seen in each one of them.

Fig. 1.14 Light line of air
and a bulk dielectric

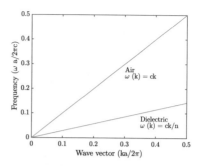

The cause of bandgap

It has been observed that a low-frequency mode tends to concentrate more of its field in the high dielectric region, while a high-frequency mode tends to concentrate more of its field in the low dielectric region, in order to maintain orthogonality between themselves (read Joannopoulos 2008 [37]). If one finds this assertion difficult to accept, he should scrutinize the light line of air and a bulk dielectric, shown in the adjacent figure. A mode of the same wave vector is of a lesser frequency in the dielectric medium than in the air (Fig. 1.14). The fact remains true for the inhomogeneous structures of the dielectric as well.

In Fig. 1.13, the modes of the first band concentrate their energy mainly in the dielectric region, and thus have lower energy and lower frequency, whereas the second band accommodates most of its energy in the air region, and hence has high energy and high frequency. As an obvious consequence, there is a substantial difference between the frequencies of the first band (a.k.a. *the dielectric band*) and the second band (a.k.a *the air band*), which manifests as the bandgap. The size of the bandgap depends directly on the dielectric contrast (i.e., difference between the high and low dielectric constants) of the photonic crystal. The greater the contrast, the broader the gap.

1.7.3   Photonic Crystal Waveguides

As mentioned above, any frequency lying in the bandgap region is forbidden from propagation. This property can be exploited to achieve confined propagation of light in a narrow channel of a homogeneous medium sandwiched between two bandgap structures. Such kind of a device is called a *photonic crystal waveguide*. Figure 1.15 illustrates the working of photonic crystal waveguides, where (a) and (b) illustrate a 1D PhC waveguide, while (c) and (d) present a 2D PhC waveguide. In the geometries shown in Fig. 1.15a and c, one can observe a narrow air region sandwiched between

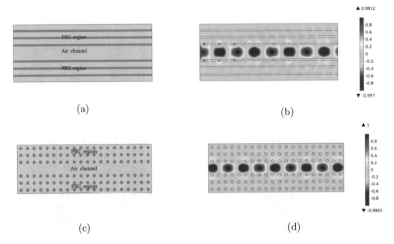

Fig. 1.15 Two types of photonic crystal waveguides—**a–b** one dimensional and **c–d** two dimensional. The blue domains in **a** and **c** indicate silicon and the gray domains represent air

the photonic bandgap (PBG) regions, where blue domains indicate silicon and gray domains represent air. In Fig. 1.15b and d, propagation of a z-polarized wave is shown where one can observe that the field remains confined in the air channel and is unable to spread in the PBG structure.

1.8 Transition from a Photonic Crystal to a Dielectric Metamaterial

During the decade 2000–2010, a variety of metallic-scatterer-based metamaterials have been developed, but they all have a common demerit called the *ohmic loss*. The ohmic loss poses a grave problem in the operation of optical metamaterials which are nanoscale in size and are meant to handle very low power. Hence, with the objective of developing a low loss alternative, all-dielectric metamaterials have become a subject of study since 2010. As a result of avid exploration, numerous all-dielectric metamaterials have been developed that are practically free from ohmic loss [27, 28, 35, 49–69]. The all-dielectric metamaterials are basically photonic crystals with certain strategic modifications, resulting in metamaterial-like behavior.

The working of a photonic crystal is based on the principle of *Bragg scattering* or diffraction. Although both refer to the same phenomenon, the term diffraction is generally used in reference to individual objects such as a slit, a grating, or a thin wire, while the term Bragg scattering is used in context to crystal lattices, being mindful of the significant contribution of the father–son duo William Bragg and Lawrence Bragg in the area [51]. Being the same as diffraction, Bragg scattering is relatively

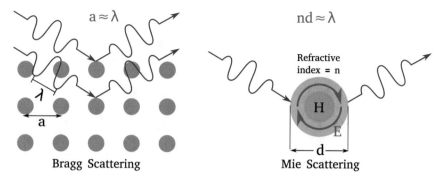

Fig. 1.16 Schematic illustration of Bragg scattering versus Mie scattering

stronger for the incident wavelengths (λ) comparable to the lattice constant (a), i.e., $\lambda_{Bragg} \approx a$ (Fig. 1.16).

On the other hand, the all-dielectric metamaterials are based on the principle of Mie scattering, whose rigorous mathematical analysis was propounded by Gustav Mie in 1908 [70]. Mie scattering is not essentially an array phenomenon and depends on the size, shape, and refractive index of the individual particles/scatterers. The wavelength for which Mie scattering is the strongest approximately equals the product of the size (d) of the particle with its refractive index (n), i.e., $\lambda_{Mie} \approx nd$. Despite a number of scatterers arranged in the form of an array of periodicity a, it is still the refractive index n and the particle size d that largely decide the scattered wavelength, while the periodicity a has negligible control over it. However, the scattered power does increase due to a large number of scatterers working together. For an array of particles of high refractive index (> 10), the scattered wavelength λ_{Mie} is large compared to the size (d). Hence, the effective medium theory [71–73] becomes applicable, and the array acts like a homogeneous slab. In this way, the structure acquires the qualities and earns the label of an *all-dielectric metamaterial*.

1.8.1 How Can a Photonic Crystal Be Made to Work as a Metamaterial?

Any array of dielectric particles exhibit both Bragg scattering and Mie scattering, but whether it shall behave predominantly as a photonic crystal or a metamaterial depends on the relative values of the two wavelengths λ_{Bragg} and λ_{Mie} [74–78]. In any periodic dielectric structure, $\lambda_{Bragg} \approx a$ and for the structure to behave as a metamaterial, $\lambda_{Mie} >> a$. This gives a relation between λ_{Bragg} and λ_{Mie} to distinguish between the two modes of operation, i.e., for a photonic crystal to behave as a metamaterial

$$\lambda_{Mie} > \lambda_{Bragg} \approx a \tag{1.62}$$

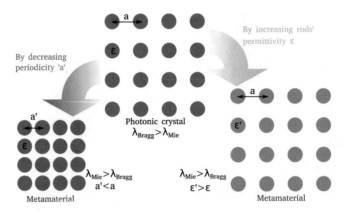

Fig. 1.17 The two routes of transition from photonic crystal to metamaterial

otherwise the photonic crystal behavior prevails. Rybin et al. proposed a phase diagram showing the transition from photonic crystal to metamaterial behavior [77], similar to the phase diagram showing the transition between three states of matter, commonly used in thermodynamics. They suggested two ways of transition from PhC to MM—(1) by reducing the lattice constant a and (2) by increasing the permittivity ϵ of the particles. Both the routes fulfill the same objective of making λ_{Mie} greater than λ_{Bragg}, hence making it greater than the lattice constant a too.

Figure 1.17 shows the two routes of transition from photonic crystal to metamaterial, for a common square array of dielectric rods-in-air-type photonic crystal. It should be noted that a sparse array of rods of low permittivity will behave as a PhC since it will have $\lambda_{Bragg} > \lambda_{Mie}$. The situation can be reversed, either by decreasing the Bragg wavelength or by increasing the Mie wavelength. The Bragg wavelength can be decreased by reducing the lattice constant, i.e., by making the array denser, while the Mie wavelength can be increased by increasing the permittivity of the rods. Increasing the rod's diameter d can also increase λ_{Mie}, but there is a limit to it, as one cannot have $d >= 0.5a$, or the structure gets transformed into its complementary version. But there is sufficient availability of high permittivity materials, especially in the low-frequency region of the spectrum.

We analyzed both the techniques numerically and the results obtained have been shown in Fig. 1.18. The red arrow tracks the Bragg scattering peak while the blue arrow points to the Mie scattering peak. Figure 1.18a–b shows the effect of variation of lattice constant, keeping the radius and permittivity of the rods constant at the values 170 nm and 12, respectively. It can be observed that as the lattice constant decreases from 1.7 to 0.378 μm, i.e., r/a increases from 0.1 to 0.45, the Bragg scattering peak undergoes a continuous blue shift, which seems obvious according to Bragg's law. According to the rods' parameters, the Mie scattering wavelength is expected to be around $\lambda_{Mie} \approx 2r\sqrt{\epsilon} = 1.177$ μm. Initially, there is no Mie peak visible between 1.0 μm and 1.5 μm (region highlighted by a dashed ellipse) in the first two graphs, but as soon as λ_{Bragg} becomes smaller than λ_{Mie}, the Mie scattering

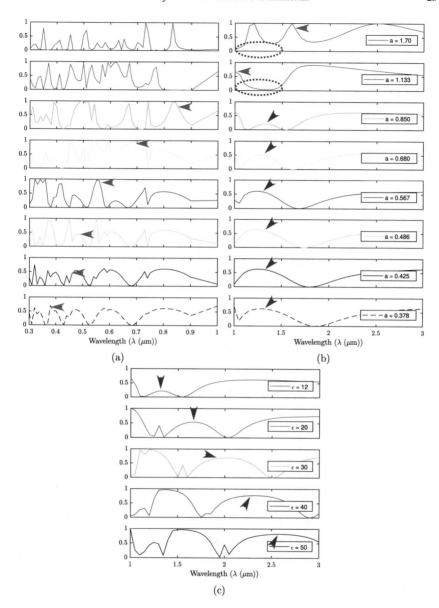

Fig. 1.18 The reflection spectrum of the array showing the effect of lattice constant and permittivity. The Bragg peak is being tracked by the red arrow while the Mie scattering peak has been indicated by the blue arrow

peak starts to appear in the said region and intensifies as the array densifies. It should be noted that with a change in periodicity (a) there is no significant shift in the position of Mie peak, attributed to the fact that it only depends on the rods' radius and permittivity, which are invariant in this case. Hence, in denser arrays where the lattice constant is substantially smaller compared to λ_{Mie}, the effective medium theory becomes applicable and the structure acquires metamaterial character.

Furthermore, Fig. 1.18c shows the effect of the second important parameter, the permittivity. Here the permittivity has been varied from 12 to 50, keeping r and a constants at 170 nm and 850 nm, respectively. It can be seen that λ_{Mie} undergoes redshift as the permittivity increases. For $\epsilon = 12$, λ_{Mie} is the same as above, which is not very large compared to $a = 850$ nm. But as ϵ increases, so does λ_{Mie}, making the structure more suitable for application of the effective medium theory and consideration as a metamaterial. In this way, it has been proved that to make the photonic crystal act as a metamaterial, either the periodicity should be reduced or the permittivity should be increased. We believe that the above discussion is sufficiently rigorous to explain the difference between a photonic crystal and a metamaterial and the transition from one to the other.

1.8.2 Photonic Crystals as Metamaterials

One of the most interesting phenomena achieved by metamaterials is the negative refractive index. Conventional design of negative-index metamaterials included metallic split-ring resonators or metallic fishnet-type structures. Contrarily, the new generation variants are purely dielectric and exhibit no power dissipation [79–84]. The simplest all-dielectric route to the negative refractive index is the abovementioned rods-in-air-type photonic crystal. Figure 1.19a shows the photonic band diagram of a square lattice of silicon rods of periodicity $a = 1\,\mu$m and radius $r = 0.25a$ and dielectric constant $\epsilon = 12.25$. By using bands 2 and 4, the refractive index has been calculated around the center of the Brillouin zone and shown in Fig. 1.19c. One can notice the spectral regions of negative- and positive-index values, as well as an empty region between them where no real value of refractive index exists. This forbidden region corresponds to the photonic bandgap between the two bands. It indicates that if a frequency belonging to the negative-index regime is allowed to travel through the photonic crystal, it will undergo negative refraction. This claim gets confirmed in Fig. 1.19e (a far-field plot), where a wave of frequency $\omega = 760 \times 10^{12}$ rad/s traveling from a prism of the photonic crystal to air emerges on the same side of the normal as the incident beam, a clear manifestation of negative refraction.

As negative index is achieved, one may wonder if it is possible to bring the negative- and positive-index curves of Fig. 1.19c close together to get a single continuous curve passing through the zero of y-axis at a particular frequency. The answer is: Yes, if one chooses a suitable radius of the rods, such that the photonic bands 2 and 4 intersect at a common point, obliterating the gap. This condition was fulfilled for radius $r = 0.2a$. Figure 1.19b shows the band diagram for $r = 0.2a$ case, in which

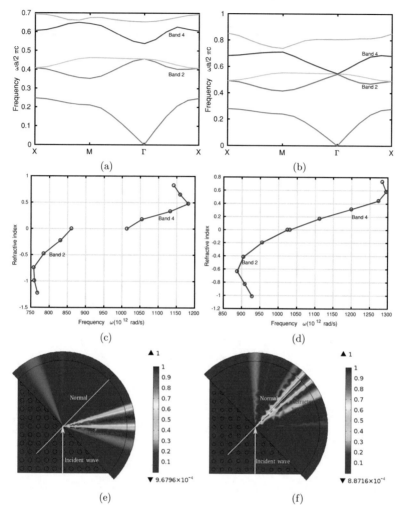

Fig. 1.19 A photonic crystal as a negative-index metamaterial (**a**), (**c**), and (**e**) and zero-index metamaterial (**b**), (**d**), and (**f**)

the bands 2 and 4 can be seen intersecting at a common point popularly known as the *Dirac point* (details in Chap. 2) [27, 60, 85]. The refractive index curve for this case has been shown in Fig. 1.19d, in which one can notice that the refractive index tends to zero at $\omega = 1025 \times 10^{12}$ rad/s. The zero-index nature gets further confirmed by zero refraction of light of frequency $\omega = 1025 \times 10^{12}$ rad/s, as shown in Fig. 1.19f, as it travels from the photonic crystal to air. It will be later explained in Chap. 2 that a wave emerging from a zero-index medium always emerges normally, irrespective of the angle of incidence. This book is dedicated to zero-index metamaterials and zero refraction property. The profound details of this category of metamaterials will be presented in the forthcoming chapters. This chapter only presents the prerequisite information needed to understand the working of a variety of zero-index metamaterials and their applications.

Chapter 2
Zero-Index Metamaterials

2.1 Introduction

As explained in the previous chapter, the way light interacts with a particular material depends on the number of electrons in it, and also on how strongly the bound electrons are held by the nucleus. For a wavelength very large compared to the lattice constant, e.g., the visible wavelength range 4000–7000 A^o compared to the lattice constant of silicon $\approx$5 A^o, different materials exhibit different effective material parameters, viz., the refractive index (n), the impedance (z), the permittivity (ϵ), and the permeability (μ) [19, 71, 72, 86–88]. Similarly, the artificial structures called *metamaterials* exhibit effective material parameters for the wavelengths very large compared to the resonator size and array periodicity. However, the advantage of metamaterial is that their effective material parameters depend more on their design peculiarities than their material. Hence, abnormal values of material parameters such as negative refractive index, negative permeability, near-zero permeability, near-zero refractive index, etc. are achievable, which is not possible with natural materials [17, 18, 27, 30, 79, 89–92]. Having near-zero material parameters is not an unprecedented phenomenon. All metals exhibit zero permittivity at the plasma frequency and a few polaritonic materials like silicon carbide (SiC) exhibit zero permeability [93, 94]. However, metamaterials have opened a way to achieve near-zero optical parameters in a more controlled and versatile manner. Depending on whether permittivity or permeability or both are zero, metamaterials can be classified into epsilon-near-zero (ENZ), mu-near-zero (MNZ), and epsilon-and-mu-near-zero (EMNZ) categories. All these metamaterials are *zero-index metamaterials* (ZIMs). However, the majority of the literature presents the definition of ZIM as the one which has both $\epsilon = 0$ and $\mu = 0$. The zero refractive index phenomenon has opened doors to numerous useful applications, which will be discussed in the later chapters of this book.

© The Author(s), under exclusive license to Springer Nature Singapore Pte Ltd. 2021
N. Shankhwar and R. K. Sinha, *Zero Index Metamaterials*,
https://doi.org/10.1007/978-981-16-0189-7_2

Fig. 2.1 Permittivity of
silicon carbide (SiC) in
mid-infrared spectrum

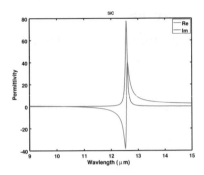

2.2 Silicon Carbide (SiC)—A Natural Zero-Index Material

Silicon carbide is a polaritonic material whose resonance falls in the mid-infrared
spectrum [93–96]. Figure 2.1 shows that silicon carbide exhibits $Re(\epsilon) = -0.0009$ ($\approx$
0) at 10.3 μm, which remains close to zero from 10.0 μm to 10.5 μm. On this basis,
it seems an excellent natural candidate for zero-index applications, except a major
challenge in the substantial value of extinction coefficient, which renders the material
significantly lossy. Besides silicon carbide, other materials exhibiting near-zero properties in the optical spectrum are aluminum zinc oxide (AZO) and indium tin oxide
(ITO), albeit the same problem of high extinction coefficient limits their performance
too [97–101].

2.3 Rectangular Waveguide—An Artificial Zero-Index
System

Though the above discussion proposes the metamaterial route to achieve zero refractive index, the "good old friend" rectangular waveguide has always been showing
this property at its cut-off frequency. Though the rectangular waveguides have been
in use for more than a century, it is just that nobody paid much attention to their
zero-index property, probably because they have always been operating above their
cut-off frequency [3, 102, 103]. For a better understanding, let us assume a rectangular waveguide like the one shown in Fig. 2.2a.

For TE$_{mn}$ mode, where the m-index corresponds to the larger dimension a and
the n-index to the smaller dimension b, the wave number k is given by [3]

$$k = \sqrt{\left(\frac{\omega}{c}\right)^2 - \pi^2\left[\left(\frac{m}{a}\right)^2 + \left(\frac{n}{b}\right)^2\right]}$$

(2.1)

$$k = \frac{1}{c}\sqrt{\omega^2 - \omega_{mn}^2}$$

(2.2)

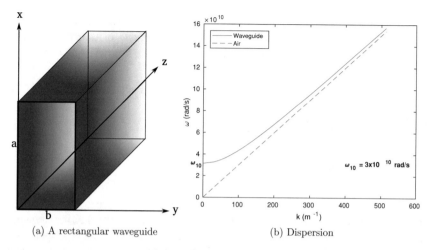

(a) A rectangular waveguide

(b) Dispersion

Fig. 2.2 **a** Design of a rectangular waveguide and **b** its dispersion curve indicating zero phase at the cutoff frequency ω_{10}, corresponding to TE$_{10}$ mode

where

$$\omega_{mn} = c\pi\sqrt{\left[\left(\frac{m}{a}\right)^2 + \left(\frac{n}{b}\right)^2\right]} \tag{2.3}$$

is the cutoff frequency. The wave velocity or the phase velocity is given as

$$v_p = \frac{\omega}{k} = \frac{c}{\sqrt{1 - (\omega_{mn}/\omega)^2}} \tag{2.4}$$

and the group velocity is given by

$$v_g = \frac{1}{dk/d\omega} = c\sqrt{1 - (\omega_{mn}/\omega)^2} \tag{2.5}$$

Below the cutoff frequency ($\omega < \omega_{mn}$), no mode exists as k is imaginary, while for very high frequencies ($\omega \gg \omega_{mn}$), propagation becomes similar to that in free space. At very high frequencies, the wavelength λ_{mn} is very small compared to the dimensions a and b so much so that the incoming wave feels as if it is traveling in free space. The lowest cutoff frequency is ω_{10}, corresponding to TE$_{10}$ mode. Figure 2.2b shows the dispersion curve for the TE$_{10}$ mode, along with the dispersion of free space (or air). It is visible that as the frequency increases, the dispersion approaches free-space behavior. The curve has been plotted for a waveguide of dimension $a = 3.0$ cm and $b = 2.0$ cm, whose lowest cutoff frequency $\omega_{10} = 3 \times 10^{10}$ rad/s. Below ω_{10} no mode exists, and all the higher modes have cutoff frequencies higher than ω_{10}.

For any mode, at $\omega = \omega_{mn}$ the wave vector k is zero. Hence, the effective refractive index is zero, while the phase velocity is infinite. Please note that no laws of physics get violated if phase velocity becomes greater than c, or even infinite, as it has no physical significance. The group velocity is the actual rate of transfer of energy, which is always less than c, according to Eq. 2.5. Hence, zero refractive index is a very natural property of certain physical systems.

2.4   Zero-Index Metamaterials

2.4.1   Basic Idea

Any medium whose effective refractive index varies from negative to positive range is bound to exhibit zero refractive index at the frequency of transition, i.e., the frequency at which the refractive index curve intersects the x-axis. This type of dispersion is abnormal for natural materials but plausible for metamaterials. For example, let us consider the well-acknowledged fishnet-type negative-index metamaterial. A fishnet metamaterial consists of a dielectric layer sandwiched between two metal layers and each layer has the same patterning throughout [31, 32, 104]. A typical unit cell of a fishnet metamaterial is shown in Fig. 2.3a and its effective refractive index is shown in Fig. 2.3b. The effective refractive index has been calculated from s-parameters using retrieval technique of Smith et al. [19], according to which the effective material parameters, i.e., the refractive index (n), the impedance (z), the permittivity (ϵ), and the permeability (μ) are calculated as follows:

$$n = \frac{1}{kd} cos^{-1} \left[\frac{1}{2S_{21}} (1 - S_{11}^2 + S_{21}^2) \right] \tag{2.6}$$

$$z = \sqrt{\frac{(1 + S_{11})^2 - S_{21}^2}{(1 - S_{11})^2 - S_{21}^2}}, \tag{2.7}$$

$$\epsilon = n/z, \tag{2.8}$$

$$\mu = nz \tag{2.9}$$

It can be seen in Fig. 2.3b that as the wavelength increases, the refractive index transits from positive to negative values, continues to remain negative for a range of wavelengths, and then returns to the positive domain. The two points of transition, corresponding to 1530 nm and 1630 nm, are the points of zero refractive index behavior. Investigation of optical properties at these two wavelengths can pave the way for several interesting applications. However, being a metallic structure, a fishnet metamaterial has huge ohmic loss, rendering it a poor choice as a zero-index metamaterial.

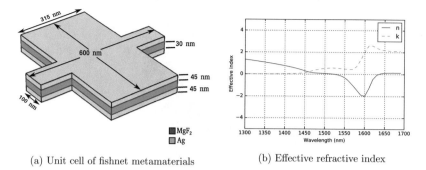

(a) Unit cell of fishnet metamaterials   (b) Effective refractive index

Fig. 2.3 Fishnet metamaterial

How to determine effective material parameters?

A metamaterial is mostly a periodic array of metallic or dielectric or composite resonators. A periodic structure exhibits periodic distribution of the fields. Hence, to simulate the operation of the metamaterial, one need not to create and computed a large array. Rather, a unit cell with appropriate boundary conditions represents the whole array, and serves the purpose of computation of the fields, s-parameters, current density, etc. The s-parameters used in Eqs. 2.6–2.7 have been computed using COMSOL Multiphysics® [47]. The technique of computation has been described below in detail.

Figure 2.4 shows the computational cell containing the unit cell of the fishnet metamaterial used for the computation of s-parameters. The dimensions and the materials of the unit cell are described in Fig. 2.3a, and the surrounding medium is air. The values such as permittivity (ϵ) for metallic and dielectric domains were adapted from Palik (1999) [105]. The top surface of the cell was made the *input port* (port 1), while the bottom surface was made the *output port* (port 2), and d is the distance between the two ports. The incident wave was excited at the port 1 and received at port 2. All the four lateral boundaries were made periodic boundaries by imposing the *periodic boundary condition* to induce a periodic array like behavior. The orientation of the electric (E) and magnetic (H) fields and the direction of propagation (k) were as shown in the figure. The computation was run in *frequency domain* for the desired range of frequencies, using the *parametric sweep* feature. As the simulation runs, several *global parameters* gets evaluated, including the *s-parameters*. Since two ports were involved, four types of s-parameters were obtained, viz., S_{11}, S_{12}, S_{21}, and S_{22}. Out of these, S_{11} and S_{21} are the most important, since they yield the reflection and the transmission coefficient of the metamaterial, respectively. Furthermore, they are also used to evaluate the *effective material parameters* (n, z, ϵ, and μ) of the metamaterial, using Eqs. 2.6–2.9.

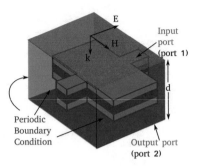

Fig. 2.4 Computational cell used for the computation of s-parameters using COMSOL Multiphysics

2.4.2   Dirac Cones—A Deep Insight into the Zero-Index Behavior

A few exotic materials have become popular topics of research nowadays, *graphene* is one of them. Graphene is a two-dimensional single-layer arrangement of carbon atoms in a hexagonal lattice (Fig. 2.5a). The exotic material has an exotic band structure having a conical-shaped feature in the valence and conduction bands, called the Dirac cones, intersecting at a common point (Fig. 2.5b) [106, 107]. Due to the conical shape of the dispersion surfaces, the dispersion in the vicinity of Dirac point is linear $E = \hbar k v_F$, where E is energy, k is wave vector, $\hbar (= h/2\pi)$ is the reduced Planck's constant, and v_F is the Fermi velocity. On account of the linear dispersion, the behavior of the conduction electrons in graphene differs from that in metals and insulators, which are parabolic dispersion materials.

In 2009, Wang et al. [85] analyzed the possibility of Dirac cones in the photonic band structure of typical optical systems and laid down the fundamental principle of

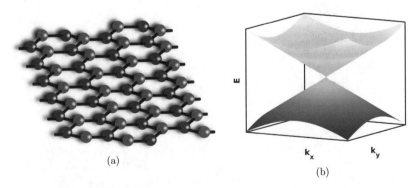

Fig. 2.5 a Structure of graphene—hexagonal lattice; **b** Dirac cones in the band structure obtained from the dispersion relation $E = \hbar k v_F$

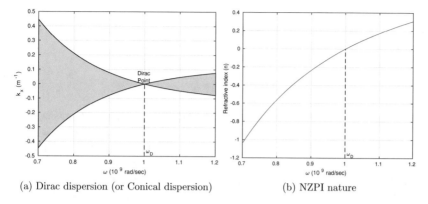

(a) Dirac dispersion (or Conical dispersion)   (b) NZPI nature

Fig. 2.6 **a** Dirac-like dispersion in a negative-zero-positive-index metamaterial (conical shape is evident) and **b** the corresponding refractive index curve

what would later cause the advent of all-dielectric zero-index metamaterials. They argued that an optical system that sustains double Dirac cones, which exhibits Dirac-like linear dispersion (see Fig. 2.6a), is ideal for the realization of zero refractive index.

Mathematically, the dispersion relation of an optical system is simply the relation between the wave vector and the frequency. One can write the wave vector k as a function of angular frequency ω, in Taylor series [108] form about the Dirac point as

$$k(\omega) = k(\omega_D) + \frac{k'(\omega_D)}{1!}(\omega - \omega_D) + \frac{k''(\omega_D)}{2!}(\omega - \omega_D)^2 + \dots \qquad (2.10)$$

where ω_D is the frequency corresponding to the Dirac point. If $k(\omega_D) = 0$ and the quadratic onward higher order terms can be ignored, then the dispersion becomes linear, and Eq. 2.10 gets reduced to

$$k(\omega) = \frac{\omega - \omega_D}{v_D} \qquad (2.11)$$

where $v_D = (d\omega/dk)|_{\omega=\omega_D} = 1/k'(\omega_D)$ is the group velocity at the Dirac point. It is visible in Eq. 2.11 that for $\omega < \omega_D$, $k(\omega) < 0$ at $\omega = \omega_D$, $k(\omega) = 0$ and for $\omega > \omega_D$, $k(\omega) > 0$. It means that the value of $k(\omega)$ varies from negative, through zero, to positive with respect to frequency. The refractive index, given by $n(\omega) = k(\omega)/k_0$, follows the same trend as k, with $n(\omega) < 0$ for $\omega < \omega_D$, $n(\omega) = 0$ at $\omega < \omega_D$, and $n(\omega) > 0$ for $\omega > \omega_D$ (see Fig. 2.6b). The authors of ref [85, 109] labeled such media as *negative-zero-positive-index* (NZPI) media and advocated the use of low loss metamaterial for the purpose.

In order to realize an NZPI medium, let us assume a metamaterial exhibiting both electric and magnetic activities and whose effective permittivity (ϵ_{eff}) and permeability (μ_{eff}) are given by Drude's model as [3, 110]

$$\epsilon_{eff}(\omega) = 1 - \frac{\omega_{ep}^2}{\omega^2 + i\omega\gamma} \tag{2.12}$$

$$\mu_{eff}(\omega) = 1 - \frac{\omega_{mp}^2}{\omega^2 + i\omega\gamma} \tag{2.13}$$

where ω_{ep} and ω_{mp} are the electric and magnetic plasma frequencies, and γ ($<< \omega_{ep}, \omega_{mp}$) is the loss factor. According to the above equations, when $\omega \to \omega_{ep}$, $\epsilon_{eff}(\omega) \to 0$ and when $\omega \to \omega_{mp}$, $\mu_{eff}(\omega) \to 0$. Now, if somehow, the magnetic plasma frequency happens to be equal to electric plasma frequency, i.e., $\omega_{mp} = \omega_{ep} = \omega_D$ and $\gamma = 0$, then for $\omega = \omega_D$, ϵ_{eff} and μ_{eff} both become zero. As a result, at the Dirac point frequency $\omega = \omega_D$, the effective refractive index $n_{eff} = \sqrt{\epsilon_{eff}}\sqrt{\mu_{eff}}$ also becomes zero. Figure 2.6 depicts this type of a system, where $\omega_{mp} = \omega_{ep} = \omega_D = 1 \times 10^9$ rad/s and $\gamma = 10^{-5}\omega_{ep}$. Figure 2.6a shows its dispersion diagram in which the Dirac cones and the Dirac point can be observed and Fig. 2.6b depicts the resultant effective refractive index which is zero at ω_D. Figure 2.7 illustrates the electric field distribution inside an NZPI medium at 0.5×10^9 rad/s ($\omega < \omega_D$), 1.0×10^9 rad/s ($\omega = \omega_D$) and 1.5×10^9 rad/s ($\omega > \omega_D$). It shows that as a wave initially propagating in air enters the NZPI medium, its wavelength is altered according to the refractive index of the medium (see Fig. 2.7). According to Eqs. 2.12 and 2.13, at $\omega = 0.5 \times 10^9$ rad/s, the medium becomes a negative-index medium with $n_{eff} = \sqrt{\epsilon_{eff}}\sqrt{\mu_{eff}} = -3.0$, thus the wavelength inside the medium becomes one-third of that in air (Fig. 2.7b). Whereas at $\omega = 1.0 \times 10^9$ rad/s (*i.e.*, $\omega = \omega_D$), the medium becomes a zero-index medium with $n_{eff} = 0$, and hence the wavelength inside the medium becomes infinite and the electric field attains a quasi-static state, i.e., the electric field remains constant throughout the medium, as shown in Fig. 2.7c. And finally, at $\omega = 0.5 \times 10^9$ rad/s, the medium acquires the positive-index character with $n_{eff} = \sqrt{\epsilon_{eff}}\sqrt{\mu_{eff}} = 0.56$, accordingly the wavelength becomes 1.78 times of that in air (Fig. 2.7d).

Please note that since these plots represent the steady-state fields obtained by frequency-domain computation, the effect of negative refractive index is not visible. If a time-domain computation method such as FDTD is employed [9, 44, 46], the wave propagation inside the medium will certainly exhibit the negative refraction.

2.5 Accidental-Degeneracy-Induced Dirac Cones in Photonic Crystals

Wang et al. presented a beautiful heuristic model for ZIM systems, but the real design of zero-index media exhibiting Dirac dispersion (or conical dispersion) was proposed by Huang et al. [27] in 2011. Huang reported that a square lattice of dielectric rods in air (see Fig. 2.8) of a particular radius and periodicity exhibits Dirac cones and linear dispersion, and hence exhibits zero refractive index at the Dirac point frequency.

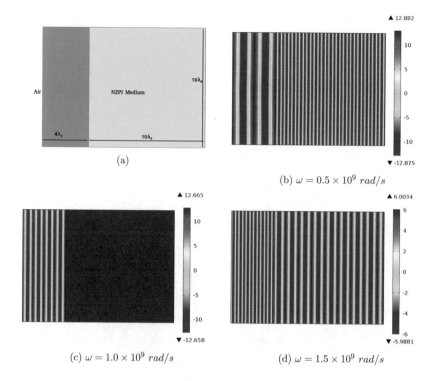

(a)

(b) $\omega = 0.5 \times 10^9 \; rad/s$

(c) $\omega = 1.0 \times 10^9 \; rad/s$

(d) $\omega = 1.5 \times 10^9 \; rad/s$

Fig. 2.7   The electric field distribution inside the NZPI medium at different frequencies

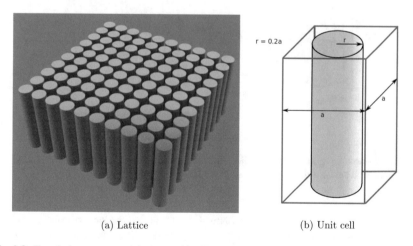

(a) Lattice

(b) Unit cell

Fig. 2.8   Zero-index metamaterial proposed by Huang et al. in 2011 [27]

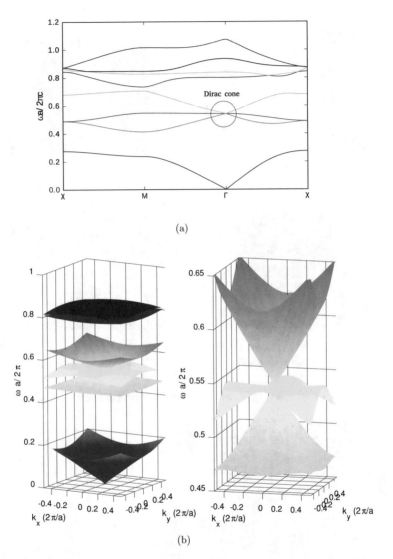

Fig. 2.9 **a** Photonic band structure for dielectric rods-in-air-type photonic crystal and **b** the equivalent dispersion surfaces illustrating the graphene-like Dirac cone

Figure 2.9a presents the photonic band structure of the photonic crystal, for which the radius-to-periodicity ratio (r/a) has been chosen as 0.2, and the relative permittivity of the dielectric has been taken as 12.5 (similar to that of silicon). One can notice a well-defined Dirac cone (highlighted by a circle) formed by the intersection of the second and the fourth bands. Moreover, the third band also passes through the intersection point, making it a triple-degenerate Dirac point. For better visualization of the conical shape, dispersion surfaces have been shown in Fig. 2.9b with a

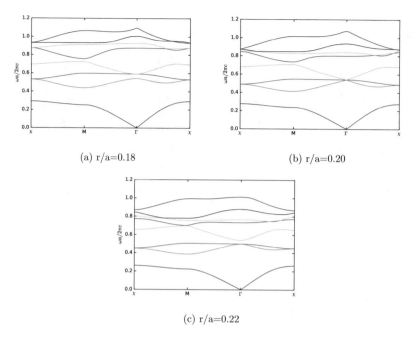

(a) r/a=0.18   (b) r/a=0.20

(c) r/a=0.22

Fig. 2.10 Role of geometrical parameters in obtaining a well-defined Dirac cone

zoomed-in view of the second, the third, and the fourth bands. The resemblance of
these double cones to those of the graphene, shown in Fig. 2.5, is noticeable, the only
difference being an extra horizontal band between them. The normalized frequency
$(\omega a/2\pi c)$ corresponding to the Dirac point is 0.541. The degeneracy here is rightly
called "accidental" because it is possible only for a particular value of radius-to-
periodicity ratio (r/a). The only way to determine the suitable value of r/a ratio is
to scan through different values and choose the best one, for which a well-defined
Dirac cone is being formed. The crucial role of r/a ratio is illustrated in Fig. 2.10.
It can be observed that the Dirac cone obtained for $r/a = 0.2$ gets ruined, even for
a slight variation of r/a to 0.18 or 0.22.

A Dirac cone is an important sign of zero-index character. However, it should
be noted that the existence of a Dirac cone is necessary but not sufficient condition
for the exhibition of the effectively zero refractive index. An additional important
factor is the electric field distribution of the three modes corresponding to the triple-
degenerate Dirac point. If the field distribution is a combination of monopole and
dipole modes, the conical dispersion can be translated to zero refractive index with
both the permittivity and permeability tending to zero at the Dirac point [27, 60,
92, 111]. Figure 2.11 shows the three types of distributions of the z-component of
electric field E_z at the Dirac point. Figure 2.11a shows the electric monopole mode,
Fig. 2.11b shows the transverse magnetic dipole mode, and Fig. 2.11c shows the
longitudinal magnetic dipole mode [70]. In light of the above observations, i.e.,

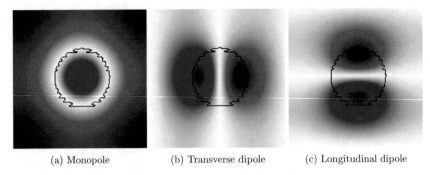

(a) Monopole   (b) Transverse dipole   (c) Longitudinal dipole

Fig. 2.11 The three types of distributions of E_z field obtained at the triple-degenerate Dirac point, i.e., at the Dirac frequency

Fig. 2.12 Effective material parameters for the square array of silicon cylinders of periodicity $a = 840$ nm and radius $r = 0.2a$

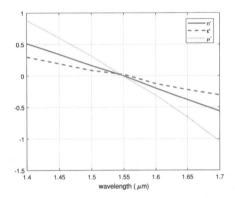

the Dirac dispersion and the monopole–dipole-combination-type field profile, it is certain that the system under consideration is a zero-index medium. Now, for the final confirmation, the effective material parameters, viz. the refractive index (n), the impedance (z), the permittivity (ϵ), and the permeability (μ), must be determined. We have considered the above metamaterial, i.e., a square array of silicon rods in air, chosen periodicity $a = 840$ nm and radius $r = 0.2a$, calculated the effective parameters using s-parameter inversion technique of Smith et al. [19]. The results of the computation are shown in Fig. 2.12.

It can be noticed that the effective relative permittivity (ϵ), relative permeability (μ), and refractive index all tend to zero at wavelength $\lambda = 1.55$ μm, or equivalently at normalized frequency $\omega a/2\pi c = a/\lambda = 0.541$, which is the same as the Dirac point frequency. Hence, the theory proposed by Wang et al. turns out to be valid and it can be stated with certainty that the Dirac cone in the band structure is indeed a signature of zero refractive index.

What about the velocity of light inside a zero-index medium?
One may wonder that zero refractive index means infinite velocity of light inside the metamaterial, which sounds unphysical. *This is a wrong perception!* One must understand that there are two types of velocities associated with electromagnetic waves: the phase velocity and the group velocity. The phase velocity is given as $v_p = \omega/k$, whereas the group velocity is given as $v_g = d\omega/dk$. The wave vector k, being zero, affects only the phase velocity and the wavelength ($\lambda = 2\pi/k$), making both of them infinite. The phase velocity is a mathematical entity, has no physical meaning, hence can be infinite. It is the group velocity that is the actual rate of transfer of energy from one point to the other inside the medium. The group velocity is calculated from the slope of the dispersion curve. For the metamaterial shown in Fig. 2.8, the group velocity at the Dirac point, calculated from the dispersion curves of Fig. 2.9a, is $0.33c$, where c is the velocity of light. Hence, **no physical law gets violated here!**

The abovementioned square lattice of rods-in-air type was the all-dielectric zero-index metamaterial [27, 60, 92]. As the research in this area progressed, several other designs for a variety of applications were developed, a few of which are shown in Fig. 2.13. Besides the square lattice of rods, there can be hexagonal/triangular or honeycomb lattices too (Fig. 2.13a, b). The rod-based structures are operational for TM polarization, while, for TE polarization, complementary structures, like holes in a dielectric slab (Fig. 2.13 c–e), have been introduced [64, 114]. The lattice of air columns can be of square, hexagonal, or honeycomb type, similar to the case of rods. In addition to these, a few exotic designs such as the star/flower-shaped holes providing high Q-factor [115] and complex designs such as honeycomb lattice of rods embedded inside a hexagonal lattice of holes have also been used for special applications like second-harmonic generation [112, 113]. Designs can be as many as a creative and innovative mind can think of, depending on the utility and ease of fabrication.

2.6 Reflection and Refraction by Zero-Index Metamaterials

When light travels from one medium to another, depending on the impedance of the two media, some part of it gets reflected, some part gets absorbed, and the rest gets transmitted [10]. The role of impedance matching in electromagnetic power transmission is well acknowledged [7, 116–119]. The greater the impedance mismatch, the greater the reflection, and the lesser the transmission. The refractive indices of the two media govern the direction of the transmitted ray with respect to incident rays, according to Snell's law of refraction. Snell's laws of refraction and reflection have been in use for a long time for systems involving natural and positive-index media, and have recently been found suitable for negative-index media as well. Zero-index media are the latest addition to this list, and in this chapter we shall see how light gets reflected or refracted when zero refractive index is involved.

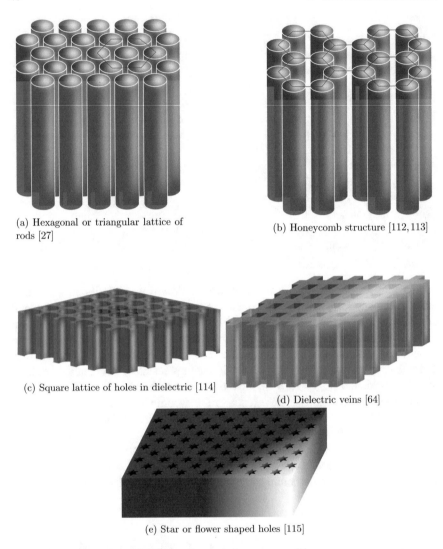

(a) Hexagonal or triangular lattice of rods [27]

(b) Honeycomb structure [112,113]

(c) Square lattice of holes in dielectric [114]

(d) Dielectric veins [64]

(e) Star or flower shaped holes [115]

Fig. 2.13 Various designs of all-dielectric zero-index metamaterials

2.6.1 Propagation from Positive Index to Zero Index

Let us consider a beam of light traveling from air to a semi-infinite slab of zero refractive index medium. The zero-index media can be of three types, viz., epsilon-near-zero (ENZ), mu-near-zero (MNZ), and epsilon-mu-near-zero (EMNZ). We know that the refractive index and the impedance of a medium are given as $n = \sqrt{\epsilon\mu}$ and $z = \sqrt{\mu/\epsilon}$, respectively. Considering *normal incidence*, for the sake of simplicity, the reflection coefficient R is given by [66] (Figs. 2.14, 2.15 and 2.16)

Fig. 2.14 High reflection in case of ENZ metamaterials

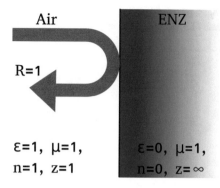

$$R = \left| \frac{z-1}{z+1} \right|^2 \tag{2.14}$$

and the transmission coefficient

$$T = 1 - R \tag{2.15}$$

Let us see how the reflection and the transmission get modified in various types of zero-index media.

2.6.1.1 Case 1: Epsilon-near-zero (ENZ)

Let us first consider the case of an ENZ metamaterial, in which $\epsilon \approx 0$ but $\mu = 1$, hence the refractive index $n = \sqrt{\epsilon\mu} = 0$ and the impedance $z = \sqrt{\mu/\epsilon} \to \infty$. Therefore, the reflection coefficient R is given by

$$R = \lim_{z \to \infty} \left| \frac{z-1}{z+1} \right|^2 \tag{2.16}$$

$$= \lim_{z \to \infty} \left| \frac{1 - \frac{1}{z}}{1 - \frac{1}{z}} \right|^2 \tag{2.17}$$

$$= \left| \frac{1 - \frac{1}{\infty}}{1 - \frac{1}{\infty}} \right|^2 \tag{2.18}$$

$$= 1 \tag{2.19}$$

The reflection coefficient R being equal to unity indicates that the light falling on an ENZ medium is 100% reflected. The deduction seems predictable since the impedance of the EMZ medium is very large. It results in an astronomical impedance

Fig. 2.15 High reflection in case of MNZ metamaterials

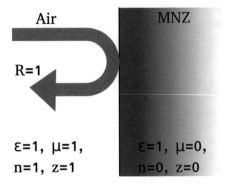

mismatch between the two media, and hence the 100% reflection. Next, we examine the case of mu-near-zero material.

2.6.1.2   Case 2: Mu-Near-Zero (MNZ)

Let us now consider the case of MNZ metamaterial, in which $\epsilon = 1$ but $\mu \approx 0$, and hence the refractive index $n = \sqrt{\epsilon\mu} = 0$, as well as the impedance $z = \sqrt{\mu/\epsilon} = 0$. Therefore, the reflection coefficient R is given by

$$R = \lim_{z \to 0} \left| \frac{z-1}{z+1} \right|^2 \tag{2.20}$$

$$= \left| \frac{0-1}{0+1} \right|^2 \tag{2.21}$$

$$= 1 \tag{2.22}$$

In the case of MNZ material too, the reflection coefficient turns out to be unity, which is understandable based on the negligible impedance of the MNZ medium. Again the impedance mismatch between air and the MNZ medium is enormous, and hence the reflection coefficient is 100%. Both the ENZ and the MNZ metamaterials are excellent reflectors. Next, we examine the reflective properties of the third candidate, the EMNZ metamaterials.

2.6.1.3   Case 3: Epsilon-mu-near-zero (EMNZ)

Finally, let us now consider the case of EMNZ metamaterial, in which both $\epsilon \approx 0$ and $\mu \approx 0$. In such a medium, the refractive index $n = \sqrt{\epsilon\mu} \approx 0$ and the impedance $z = \sqrt{\mu/\epsilon} \neq 0, \neq \infty$. Here the impedance has a finite value, which can be brought close to unity, if relative permittivity can be made approximately equal to relative permeability, i.e., ($\epsilon \approx \mu$). Then, the impedance of the EMNZ medium becomes

Fig. 2.16 High transmission in case of EMNZ metamaterials

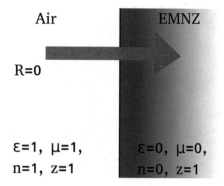

close to that of air, and the reflection coefficient R becomes

$$R = \left| \frac{z-1}{z+1} \right|^2 \tag{2.23}$$

$$\approx \left| \frac{1-1}{1+1} \right|^2 \tag{2.24}$$

$$\approx 0 \tag{2.25}$$

Zero reflection coefficient means almost 100% transmission, i.e., almost all the light falling on the EMNZ metamaterial gets transmitted with negligible reflection. Such a characteristic is desirable, and the metamaterial is so strategically designed that ϵ and μ both tend to zero at the same wavelength. Alternatively, if $\mu \neq \epsilon$, the reflection coefficient falls between 0 and 1 ($0 < R < 1$).

2.6.2 Role of the Angle of Incidence

Above we considered normal incidence for the sake of simplicity, but it is worthwhile to study the case of oblique incidence for the significance of generality. We saw above that when light falls normally on a zero-index medium from the air, most of it gets transmitted to the ZIM, and the angle of refraction is also bound to be almost zero in accordance with Snell's law.

$$n_1 sin\, \theta_i = n_2 sin\, \theta_r \tag{2.26}$$

where n_1 and θ_i are the refractive index of medium 1 (air in this case) and the angle of incidence, while n_2 and θ_r are the refractive index of medium 2 (ZIM in this case) and the angle of refraction, respectively. According to Snell's law, if $n_1 = 1$ and $\theta_i = 0$, then θ_r will also be zero irrespective of n_2, which holds good in the case zero-index medium also.

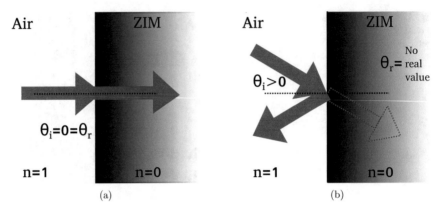

Fig. 2.17 Fate of light falling on a zero-index metamaterial slab, according to Snell's law, for **a** normal incidence **b** oblique incidence

Figure 2.17 shows the case of normal incidence. We see that light incidents normally on the slab and propagates normally inside it. This behavior has been verified numerically as well, and the result of the simulation has been presented in Fig. 2.18a. It can be seen that light penetrates, as it is, into the zero-index region. However, an interesting behavior is observed when light falls obliquely on the ZIM, i.e., at any angle greater than zero. By pondering a little, one can easily figure out the reason for this behavior. Since $n_2 \approx 0$, the critical angle for the system shown in Fig. 2.17 is $\theta_c = sin^{-1} (n_2/n_1) \approx 0$. Thus, light is able to propagate into the zero-index medium only if $\theta_i \approx 0$ and for any value of θ_i substantially larger than zero, it gets total internally reflected. Mathematically, if $\theta_i >> 0$ and $n_2 \approx 0$, then $sin \theta_r = n_1 sin \theta_i/n_2 >> 1$, therefore θ_r cannot acquire any real value. Consequently, no refraction but only total internal reflection is feasible. This prediction by Snell's law has been schematically shown in Fig. 2.17b and numerically confirmed and demonstrated in Fig. 2.18b. This property of total internal reflection is very interesting and can be very useful in making on-chip waveguides with zero-index cladding. An interesting phenomenon attached to the total internal reflection is Goos–Hänchen shift, attributed to the slight penetration of light into the cladding. The Goos–Hänchen shift has been deeply studied for significant optical waveguidance systems like optical fibers and planar waveguides, in order to find ways to minimize it. Moreover, a negative Goos–Hänchen shift has been observed and studied in the negative-index-metamaterial-based waveguidance systems. Hence, it is worthwhile to investigate the modification of the GH shift phenomenon in zero-index-metamaterial-based devices too. In fact, in Chap. 3, we have rigorously analyzed the Goos–Hänchen shift for a glass–ENZ system for both s- and p-polarizations. That is all worth mentioning about the propagation of light from a positive to a zero-index medium. Next, we consider the reverse case, i.e., light propagating from a zero-index medium to air.

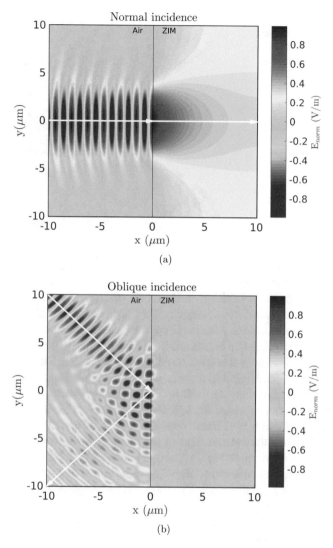

Fig. 2.18 Numerical analysis of light traveling from air to ZIM at **a** normal and **b** oblique incidence

2.6.3 Propagation from Zero Index to Positive Index

Let us now consider the reverse case than that in the previous section, and assume light to be propagating from a zero-index medium to air. Now, n_1 (≈ 0) is the refractive index of ZIM, and n_2 ($= 1$) is the refractive index of air. According to Snell's law (Eq. 2.26), as light travels from ZIM to air, the angle of refraction $\theta_r = sin^{-1}(n_1 sin\ \theta_i / n_2)\ \approx 0$, irrespective of the value of θ_i, since the $n_1 \approx 0$. In

Fig. 2.19 Schematic
illustration of the zero-index
prism used to analyze the
propagation of light from a
zero-index medium to air

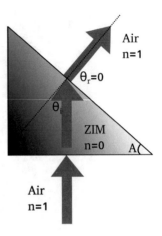

simpler words, light always emerges normally from a zero-index medium, irrespective of the angle of incidence.

In this case too, we numerically analyzed the propagation of light from ZIM to air for different angles of incidence, viz., 30°, 45°, and 90°, and found that light emerged normally out of the ZIM in all the three cases. For this purpose, we have used a right-angled prism of a homogeneous zero-index material whose second base angle A has been made equal to the desired angle of incidence. The design has been shown in Fig. 2.19. The light is fed normally into the prism from the base so that it subtends angle $\theta_i = A$ at the ZIM–air boundary (the hypotenuse) w.r.t. the normal (the dashed line). θ_i was varied through the above three values by varying A. The results of the simulation have been shown in Fig. 2.20. The left-hand side of each sub-figure illustrates the electric field of the incident wave, while the right-hand side shows that of the refracting (or emerging) wave in the far-field domain. The reason of using far-field domain to show the refraction is that the beam-like shape becomes visible and the direction of propagation and angle of refraction are more recognizable. It is visible in each case that most of the field is concentrated around the normal, confirming the normal emergence. Here we wish to mention an interesting application based on the property of normal emergence, viz., *beam steering*. Please note that by changing A, the slope of hypotenuse changes, and so does the direction of the emerging beam w.r.t. to the horizontal direction. In this way by controlling the angle A of the ZIM prism, one can control the direction of the beam emerging from it.

It should be noted here that though we have analyzed and shown the emergence of light from the hypotenuse only, the radiation comes out from the vertical side as well. The electric field is uniform throughout the ZIM and permeates all the space inside it. All the points inside the prism and at the two remaining boundaries vibrate in the same phase and magnitude as the base (i.e., the entry boundary). We have illustrated this property in Fig. 2.21 using a homogeneous block of zero-index material. Light enters the block from one side and emerges from all the remaining sides.

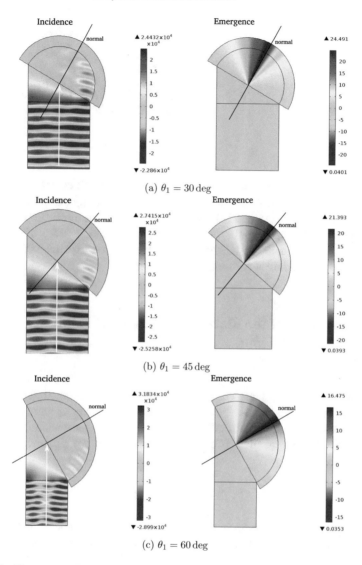

(a) $\theta_1 = 30 \deg$

(b) $\theta_1 = 45 \deg$

(c) $\theta_1 = 60 \deg$

Fig. 2.20 Almost normal emergence irrespective of angle of incidence. Left-hand side: incident wave propagating toward the ZIM–air boundary for **a** $\theta_1 = 30$ deg, **b** $\theta_1 = 45$ deg, and **c** $\theta_1 = 60$ deg. Right-hand side: Refracted beam emerging from the ZIM–air boundary. In all the three cases, $\theta_2 \approx 0$.

Fig. 2.21 Schematics of
ZIM prism used to analyze
the propagation of light from
ZIM to air

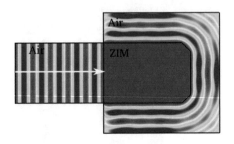

In Fig. 2.21, it has been shown that light propagating in air enters the ZIM from the left-hand side, uniformly distributes its electric field throughout (as no crests and troughs can bee seen inside the ZIM), and emerges from all the remaining boundaries. This property of ZIM can be put to numerous useful applications, a few of which have been described in Chap. 3. By this, we believe that we have explained the reflection and refraction properties of zero-index metamaterials.

2.7 Decoupling of the Electric and Magnetic Fields in Zero-Index Medium

As discussed in Chap. 1, by 1931, electricity and magnetism had been established as the two sides of the same coin. The next major leap in electrodynamics was made by Maxwell, who argued that light was an electromagnetic wave comprising of electric and magnetic field oscillations, which are the cause and effect of each other. Maxwell's curl equations in the time-harmonic form are written as

$$\nabla \times \mathbf{E} = i\omega\mu\mathbf{H} \qquad (2.27)$$

$$\nabla \times \mathbf{H} = -i\omega\epsilon\mathbf{E} \qquad (2.28)$$

In any medium, as light propagates, the electric and the magnetic fields remain interwoven with each other. However, there is a subtle way of decoupling the two seemingly inseparable entities, without violating the laws of physics, by introducing a material with near-zero refractive index, as demonstrated below [24, 91, 120]. The Eq. 2.27 can be expanded as

$$\begin{bmatrix} \hat{i} & \hat{j} & \hat{k} \\ \frac{\partial}{\partial x} & \frac{\partial}{\partial y} & \frac{\partial}{\partial z} \\ E_x & E_y & E_z \end{bmatrix} = i\omega\mu(H_x\hat{i} + H_y\hat{j} + H_z\hat{k}) \qquad (2.29)$$

For an E_x–H_y–k_z mode, the above equation gets reduced to

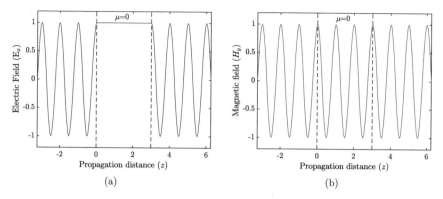

Fig. 2.22 Nature of **a** electric and **b** magnetic fields inside a mu-near-zero ($\mu \approx 0$) medium

$$\begin{bmatrix} \hat{i} & \hat{j} & \hat{k} \\ \frac{\partial}{\partial x} & \frac{\partial}{\partial y} & \frac{\partial}{\partial z} \\ E_x & 0 & 0 \end{bmatrix} = i\omega\mu H_y \hat{j} \tag{2.30}$$

or

$$\frac{\partial E_x}{\partial z} = i\omega\mu H_y \tag{2.31}$$

Now, for an MNZ medium (with $\mu \approx 0$)

$$\frac{\partial E_x}{\partial z} \approx 0 \tag{2.32}$$

From which it can be inferred that the electric field is spatially invariant (i.e., invariant with respect to z) inside a mu-near-zero medium. Please note that near-zero permeability does not affect the variation of the magnetic field H, which remains sinusoidal as usual. Figure 2.22 shows that the variation of the electric and magnetic fields inside a medium having $\mu \approx 0$.

It can be noticed that the magnitude of the electric field (blue) remains constant throughout the MNZ medium, while the magnetic field component (red) exhibits sinusoidal variation. Moreover, the phase of the electric field at the emergence boundary is the same as that at the incidence boundary. This is evidence of the fact that using an MNZ metamaterial the electric field of an electromagnetic wave can be handled independently without disturbing the magnetic field component.

On performing a similar analysis using Eq. 2.28, it can be shown that for an E_x–H_y–k_z mode in an ENZ ($\epsilon \approx 0$) medium, the magnetic field becomes uniform, while the electric field remains sinusoidal. The corresponding plots for the variation of the electric (blue) and magnetic fields (red) are shown in Fig. 2.23.

$$\frac{\partial H_y}{\partial z} = 0 \tag{2.33}$$

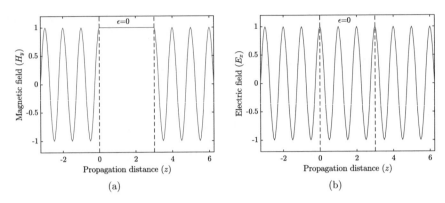

Fig. 2.23 Nature of **a** magnetic and **b** electric fields inside an epsilon-near-zero ($\epsilon \approx 0$) medium

Table 2.1 Magnitude of the electric field inside a zero-index medium, at different instants of time, according to Fig. 2.24, indicating a sinusoidal variation

Time	t = 0.0T	t = 0.2T	t = 0.4T	t = 0.6T	t = 0.8T	t = T
E_x	1.0	0.30902	−0.80902	−0.80902	0.30902	1.0

In light of the above, it is easy to conclude that inside an EMNZ metamaterial, in which ϵ and μ both tend to zero, the electric and the magnetic fields both become uniform. It is important to mention here that zero refractive index affects only the spatial variation of the fields, while the temporal variation remains unaltered and sinusoidal, as usual. Figure 2.24 shows the snapshots of the electric field distribution inside and outside a mu-near-zero medium at different instants during one cycle of oscillation. It can be seen that, though the field inside the zero-index medium is constant with respect to space, it still exhibits sinusoidal variation with respect to time. It is an important aspect of zero-index photonics, which should always be kept in mind (Table 2.1).

2.7.1   Experimental Verification of Zero Refractive Index

Although the abovementioned numerical results showing the existence of zero-index media are convincing, an experimental verification has unparalleled significance. It has already been asserted and numerically illustrated in the previous section that light does not undergo a change in phase as it travels through a zero-index medium. Based on this property Reshef et al., in 2017, demonstrated a *phase-free propagation* at the wavelength of 1627 nm, using a zero-index waveguide [121]. The waveguide used by them was a silicon slab with air holes in it. To experimentally verify the zero-index character of the waveguide, they employed a simple and unique method of *standing waves* [122, 123]. A standing wave, as it is well acknowledged, is formed when two

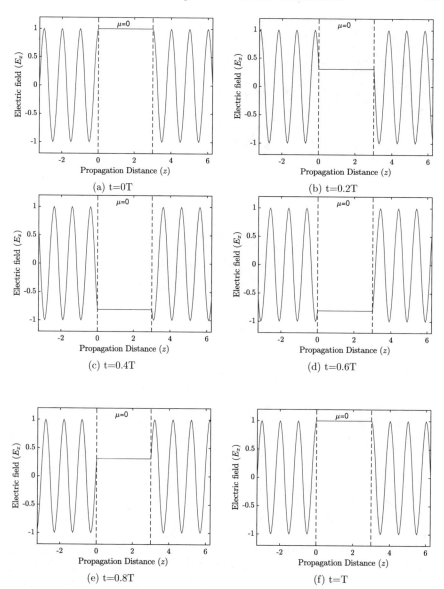

Fig. 2.24 The variation of electric field inside a mu-near-zero ($\mu \approx 0$) medium at different instants of time, illustrating the sinusoidal behavior in time domain

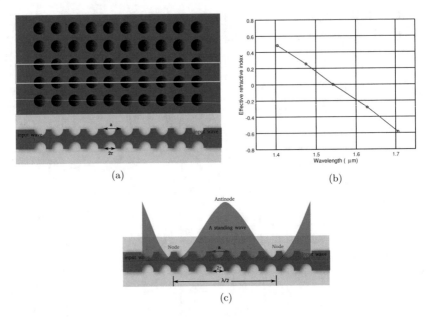

(a)   (b)

(c)

Fig. 2.25 Design and principle of phase-free propagation

waves of the same frequency traveling in opposite direction superimpose on each
other. The energy distribution of a standing wave is such that particular points in the
medium always remain at rest or vibrate the least, called the *nodes*, while certain
others experience maximum displacement or vibration, called the *antinodes*. The
vibration at the nodes can be reduced to zero if the two interfering waves are of
equal amplitude. In a waveguide of zero refractive index, the effective wavelength
must be infinite and the nodes must disappear since separation between them also
becomes infinite. The disappearance of nodes was observed by Reshef et al. during
their experiment and reported in their paper.

To numerically demonstrate the phase-free propagation, we used a similar zero-
index waveguide whose design is shown in Fig. 2.25a. The waveguide has been
designed to operate at 1550 nm (telecom wavelength) [11, 124–127]. A small strip
cut out from a zero-index metamaterial, highlighted by a green rectangle, has been
used as the waveguide here. The radius of air holes is $r = 183$ nm and the periodicity
of the holes is $a = 581$ nm. The effective refractive index of the metamaterial, as
a function of wavelength, has been shown in Fig. 2.25b. It can be observed that the
refractive index tends to zero around 1550 nm. It means that to the light of wavelength
1550 nm, the waveguide should offer zero refractive index and facilitate a phase-free
propagation.

To generate a standing wave in the waveguide, two waves of the same free-
space wavelength and amplitude were fed into the two ends. As the two antiparallel
waves superimpose, a standing wave (shown in Fig. 2.25c) is generated with nodal
and antinodal regions. It is a known fact that distance between two adjacent nodes

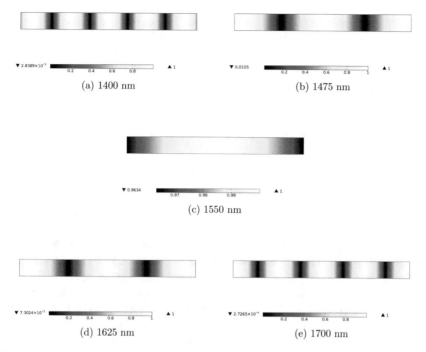

Fig. 2.26 Formation of standing waves of different wavelengths. The distance between the nodes increased as zero-index wavelength is approached

is equal to half of the wavelength inside the medium, which itself depends on the effective refractive index of the medium. Conclusively, the lesser the refractive index of the medium, the longer the effective wavelength and more distant the nodes are.

To recreate the experimental results numerically, here we have used a rectangular waveguide whose refractive index varies w.r.t. wavelength in the same manner, as it is shown in Fig. 2.25b. On launching the waves of five different free-space wavelengths into the two ends of the waveguide, five types of standing wave patterns, shown in Fig. 2.26a–e, have been obtained. The wavelength has been varied from 1400 to 1700 nm, and correspondingly the refractive index transits from positive to negative values through a zero value at 1550 nm. In Fig. 2.26, the black regions in the waveguide are the nodes, and the white regions are the antinodes. It can be observed that as the refractive index decreases, the separation between the nodes increases. The nodes of 1475 nm wave are more distant than the nodes of 1400 nm. The most interesting one is the case of 1550 nm (Fig. 2.26c), where no nodes are present in the waveguide since the effective refractive index has reduced to almost zero, and the effective wavelength has become very large. It is necessary to scrutinize the color bar in this case, because one may misinterpret the dark ends of the waveguide as nodes. Please note that the minimum of the color bar of 1550 nm is 0.9634, unlike other cases. It means that the magnitude of the field is 0.9643 at the two ends of the waveguide, which is a slight reduction than the maximum value "1" at the center, but not a node. It is a manifestation of zero-index nature, that is, as the wavelength becomes very large, the field inside the device remains almost uniform. Whereas, in all other cases, the black regions mark the actual nodes. Now, as one moves beyond 1550–1625 nm and 1700 nm wavelength, nodes reappear in the device. As the refractive index increases (though in a negative scale), the inter-nodal separation decreases, due to the shortening of the effective wavelength. It is a simplistic yet beautiful method to illustrate the zero-index nature of the metamaterial and also put forward phase-free propagation as a useful application of the zero-index phenomenon.

2.8   Tunability of a Zero-Index Metamaterial

The position of the Dirac cone, and hence the corresponding zero-index frequency depends on the geometrical and the material parameters. As a result, there can be two types of tuning of the metamaterial structure—*static tuning* and *dynamic tuning*.

Let us take the example of the commonly used square lattice of rods. The static tuning of the structure can be done by changing the diameter and pitch of the rods, and/or by changing the material of the rods. The explanation about this tunability lies in the scalability of Maxwell's equation [37]. From our knowledge about the photonic bandgap materials, we know that, in the case of a square array of dielectric columns, if the radius of the columns and their periodicity is doubled (or halved), without changing the r/a ratio, the corresponding bandgap wavelengths will be doubled (or halved) as well. The same is true for zero-index metamaterials and the corresponding zero-index wavelength, provided the formation of Dirac cone is not compromised with scaling. This type of tuning is called *static tuning*. A drawback of this method is that one needs to fabricate different structures for different wavelengths of operation (i.e., the zero-index wavelengths). As once fabricated, a structure can provide zero-index only at a single wavelength, and the index remains close to zero within a very narrow spectral region around it. A more convenient method of tuning the metamaterial to a different operational frequency would be by modifying the surrounding conditions without altering its geometry. Such tuning is called *dynamic tuning*. However, to the best of our knowledge, no such technique has been reported until the drafting of this book. Nonetheless, we have discussed below two probable routes of achieving dynamic tunability.

One may argue that by using liquids ($n > 1$) as surrounding media which can be removed or replaced as per desire (Fig. 2.27a), a dynamic control over the wavelength of operation can be achieved. We examined the feasibility of this technique numerically and the results have been shown in Fig. 2.27c–d. The results show that dynamic tunability is not easy to achieve. Figure 2.27c shows the required value r/a to achieve a well-defined Dirac cone (red curve) and the corresponding Dirac point frequency (a/λ) of the Dirac cone thus obtained (blue curve), with respect to the refractive index (n_s) of the surrounding medium. It implies that if the metamaterial is submerged in a liquid medium, it cannot immediately start operating at a different wavelength. Change in the refractive index of the surrounding calls for the change

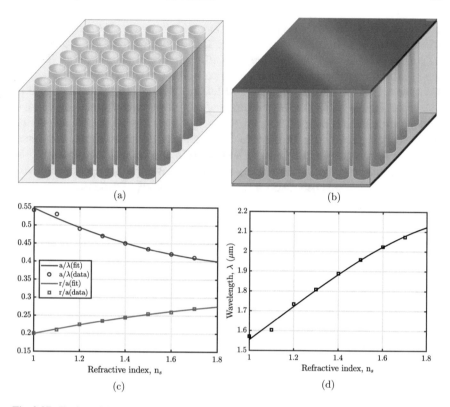

Fig. 2.27 Tuning of the zero-index metamaterial by changing the surrounding medium, and by the electro-optic effect. **a** ZIM embedded in a low-index material like silica; **b** ZIM embedded in an electro-optic material and provided with parallel plate electrodes for applying external electric field; **c** the required r/a ratio (red) for different values of the refractive index (n_s) of the surrounding medium to achieve a well-defined Dirac cone, and the normalized frequency position (a/λ) of the achieved Dirac cone (blue); **d** zero-index wavelength w.r.t. the refractive index (n_s) of the surrounding medium, for $a = 840$ nm

in the geometry of the structure (i.e., r/a ratio) as well to obtain a Dirac cone and make the metamaterial functional at a new wavelength [65]. Figure 2.27d shows the zero-index wavelength versus the surrounding refractive index (n_s). This has been obtained from the blue curve of Fig. 2.27c by transforming the normalized frequency (a/λ) to wavelength (λ) assuming $a = 840$ nm. Though the structure is being tuned from 1550 to 2100 nm, it cannot be termed dynamic tunability, since it required change in the radius r of the columns.

However, a probable technique of achieving dynamic tunability can be by using the *electro-optic effect* [11]. One can choose a material whose refractive index can be varied by application of a high electric field. Such materials are called *electro-optic materials*. Figure 2.27b shows the metamaterial embedded inside an electro-optic material, such as lithium niobate ($LiNbO_3$) [128–130], provided with metallic plates serving as electrodes for the application of electric field. The refractive index of the electro-optic material depends on the applied electric field. One can advance a step ahead by making the rods too, of a different electro-optic material. Though we have neither experimentally nor numerically analyzed the electro-optic design, we believe that the idea is fascinating and worth investigating. The advantage of dynamic methods is that there will be no need of making several structures, because a single structure can be used for a broad range of wavelengths, merely by altering the surrounding conditions. In all the tuning techniques discussed above, a universal rule must always be satisfied, i.e., *a well-defined Dirac cone must be obtained in the band structure*. If such techniques are successfully developed and mastered, it will be very beneficial and cost-effective for integrated optics applications.

Chapter 3
Applications of Zero-Index Metamaterials

3.1 Introduction

Exotic phenomena always have enchanting applications, so is the case with zero refractive index. Zero-index metamaterials have several interesting applications such as wavefront engineering, electromagnetic tunneling, electromagnetic cloaking, directivity enhancement by manipulation of the radiation pattern, etc. which have been discussed in this chapter.

3.2 Electromagnetic Tunneling

In the previous chapter, it has been shown that there is no change in the phase of the wave as it passes through a zero-index medium. The wave exits the ZIM with the same phase as with which it enters, and the magnitude of the field remains the same throughout the medium. This property can be exploited to achieve electromagnetic tunneling through distorted channels. Distortions can be of several types, such as shrinking, expansion, irregular shape, bending, etc. The tunneling phenomenon has been discussed in detail, in the following.

3.2.1 Tunneling Using Epsilon-Near-Zero (ENZ) Materials

Suppose, in a photonic integrated circuit, light is required to travel from a broad waveguide to a very narrow channel of subwavelength thickness. The impedance mismatch between the waveguide and the narrow channel will be drastic. Therefore, the reflection loss will be high, and coupling will be poor. However, if the narrow channel is filled with an epsilon-near-zero material, the incident electromagnetic wave gets squeezed into the channel and is almost entirely transmitted to the other side. In other words, the incoming radiation is capable of getting tunneled through a narrow channel of the subwavelength thickness, if the channel is filled with an epsilon-near-zero material, irrespective of how thin or irregularly shaped the channel may be (see Fig. 3.1) [91, 131].

Silveirinha et al. have presented a rigorous analysis of this phenomenon and have formulated a simple yet accurate relation for reflection coefficient of an electromagnetic wave impinging on an epsilon-near-zero-material-filled narrow channel, which is given by

$$\rho = \frac{(a_1 - a_2) + ik_0\mu_{r,p}A_p}{(a_1 + a_2) - ik_0\mu_{r,p}A_p} \tag{3.1}$$

where a_1 and a_2 are the widths of the waveguide on the two sides of the ENZ-filled distorted channel (as shown in Fig. 3.1a), k_0 is the free-space wave vector of the impinging wave, $\mu_{r,p}$ is the permeability of the narrow channel, and $A_p = wt$ is its area of cross section. It is understood that to maximize the tunneling (or transmission), the reflection coefficient ρ should be minimized. In Eq. 3.1, it has been assumed that the imaginary parts are very small compared to the real parts, i.e., $(k_0\mu_{r,p}A_p)/(a_1 + a_2) << 1$, then the reflection coefficient gets reduced to its minimum value given by

$$\rho = \frac{a_1 - a_2}{a_1 + a_2} \tag{3.2}$$

or

$$|\rho| = \frac{|a_1 - a_2|}{a_1 + a_2} \tag{3.3}$$

According to Eq. 3.3, the reflection coefficient can be suppressed to almost zero by making $a_1 \approx a_2$, thereby achieving 100% transmission. In the numerical analysis presented here, $a_1 = a_2 = 2\lambda$ has been assumed. It is important to note that such easy control over ρ in terms of geometrical parameters a_1 & a_2 could be possible because of the condition $(k_0\mu_{r,p}A_p)/(a_1 + a_2) << 1$. The condition can be satisfied in two ways—(1) if the permeability of the ENZ region is close to zero ($\mu_{r,p} \approx 0$) and (2) if the area of cross section (A_p) of the ENZ channel is very small, which is already true since the channel is narrow. Now, to examine qualitatively how small the area A_p should be, we deduce that for a non-magnetic medium, the above condition gets reduced to

$$\frac{k_0 A_p}{2a} >> 1 \tag{3.4}$$

$$\implies \frac{A_p}{2a} >> 1 \implies A_p << \frac{\lambda_0 a}{\pi} \tag{3.5}$$

or simply

$$A_p << \lambda_0 a \tag{3.6}$$

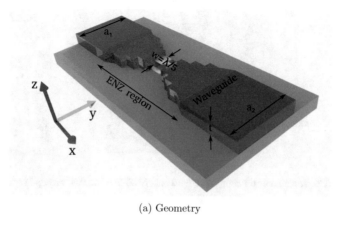

(a) Geometry

▼ -6.246   ![scale](-6 -4 -2 0 2 4 6)   ▲ 6.2406

(b) Magnetic field H_z

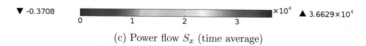

▼ -0.3708   ![scale](0 1 2 3)×10^4 ▲ 3.6629×10^4

(c) Power flow S_x (time average)

Fig. 3.1 Electromagnetic tunneling through an ENZ-filled narrow waveguide of subwavelength thickness

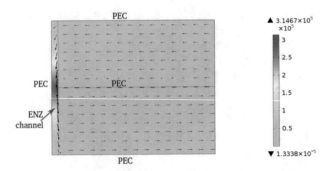

Fig. 3.2 Electromagnetic tunneling through a bent ENZ-filled narrow waveguide of subwavelength thickness

In light of the above conditions, one can achieve full transmission in the case of narrow ENZ channels.

Besides thickness, another significant deformity is the bending of the channel. An optical signal passing through a sharp bend undergoes substantial power loss. A sharp bend in the connecting lines of an optical integrated circuit can pose a severe challenge to the performance of the device, albeit they are sometimes necessary for the compactness of the circuit. Hence, a method is needed to reduce the power loss at the bends. ENZ medium proves to be a good solution to this problem too. As we saw above that an ENZ medium allows electromagnetic tunneling through a narrow channel irrespective of its shape and size, the phenomenon extends equally well to a bent waveguide too. Figure 3.2 shows the tunneling through a bent (90°) ENZ channel of thickness $\lambda/10$. The contour plot shows the norm of the electric field and the arrows illustrate the power flow. It can be observed that the power flows from the top waveguide to the bottom waveguide through the bent ENZ channel. This illustrates that an ENZ-filled channel is capable of making light propagate through bends as sharp as 90°, without significant bending loss, by electromagnetic tunneling. Marcos et al. coined a term *supercoupling* for this phenomenon. The next subsection discusses the supercoupling phenomenon using mu-near-zero (MNZ) metamaterials.

3.2.2 Tunneling Using Mu-Near-Zero (MNZ) Materials

In analogy to the epsilon-near-zero-material-filled narrow channel, supercoupling can also be achieved in a highly broadened channel that is filled with mu-near-zero (MNZ) material. Marcos et al. have discussed the MNZ supercouple in their article published in 2015 [91, 132]. Figure 3.3 illustrates the case of a widened channel filled with mu-near-zero material. In Fig. 3.3, the geometry of the structure is shown in which the width of the channel in the middle of the MNZ region is five times the wavelength, i.e., the middle region is substantially broader compared to the rest of the waveguide. The results of numerical computations shown in Fig. 3.4b, c clearly

Fig. 3.3 Schematic illustration of the geometry of an MNZ-filled distortion

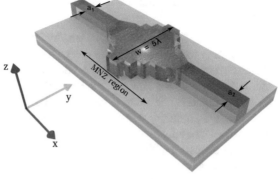

Fig. 3.4 Electromagnetic tunneling through an MNZ-filled broad waveguide of subwavelength thickness

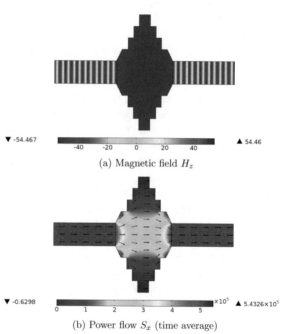

(a) Magnetic field H_z

(b) Power flow S_x (time average)

illustrate the full tunneling through the distorted region. Figure 3.4b shows the z-component of the magnetic field, and Fig. 3.4c shows the power flow. From both these figures, we observed the electromagnetic wave impinging on the MNZ from one waveguide gets transmitted, as it is, to the other side into the second waveguide. It should be understood that it is easy to achieve epsilon-near-zero naturally by simply using metals in their pure or diluted (with a dielectric) form near the plasma frequency. However, magnetic activity is difficult to achieve naturally and requires the employment of magnetic metamaterials. Marcos et al. suggested the use of split-ring resonator (SRR)-based magnetic metamaterial for this purpose.

3.2.3   Controlling the Tunneling Using Obstacles

Nguyen et al. [133] reported in 2010 that by including defects in zero-index meta-materials, their tunneling capability can be so efficiently controlled that total trans-mission or total reflection can be achieved. They used cylindrical defects of different radii to demonstrate the effect of geometrical and material parameters of the obsta-cles on the tunneling characteristics. Figure 3.5 shows how the transmission can be controlled by introducing defects inside an EMNZ medium, which has both ϵ and μ close to zero. The equal values of ϵ and μ have been assumed so that their ratio, and hence the impedance of the EMNZ medium is equal to 1. The medium on both sides of the EMNZ medium has been chosen to be air. Such a medium is called *matched-impedance zero-index material* (MIZIM). Figure 3.5a–c shows the magnetic field H_z, the electric field E_y, and the Poynting vector S_x, respectively, for an incident wave of frequency 15 THz. It can be observed that the field on the right-hand side of the EMNZ medium is substantially lower compared to the left-hand side, which indicates reduced transmission. There are three different defect objects of radii 4 μm, 8 μm, and 12.4 μm and permittivity 3.66, 11.86, and 15.67, respectively. In the absence of these defects, the wave is fully transmitted to the other side of the EMNZ slab.

The defect's size and the defect's material play a crucial role in deciding the transmission. Here, the effect of both these quantities has been numerically analyzed and the results have been presented in Fig. 3.6. Figure 3.6a shows the design of the computational region, in which the air, the EMNZ, and the defect regions have been depicted. The defect object is cylindrical and has been arbitrarily placed inside the EMNZ medium. The radius and the permittivity of the defect have been varied and the corresponding transmission coefficient has been calculated. The plots thus obtained have been shown in Fig. 3.6b, c, from which it can be inferred that for certain values of R and ϵ the transmission can be as high as 100% while for certain other it can be reduced down to almost zero. Thus, we get that the transmission characteristic of a zero-index medium can be arbitrarily controlled by introduction defects of strategically chosen geometrical and material properties.

3.3   Electromagnetic Cloaking

Electromagnetic cloaking is one of the most fantastic phenomena achieved by meta-materials, in which a metamaterial cloak manipulates the path of light around the cloaked object in such a way that the wavefront emerging out of the cloak reforms into its original shape, thereby *casting no shadow*. Incident wave completely ignores an optically large object and passes unaltered, if the object is cloaked [27, 134–147]. This application is a major reason for the popularity of metamaterial, besides neg-ative refraction, as it seems straight out of science fiction. There are two ways to electromagnetic cloaking: (1) by means of transformation optics; (2) by means of zero refractive index.

Fig. 3.5 Controlling the transmission through EMNZ medium by inserting defects

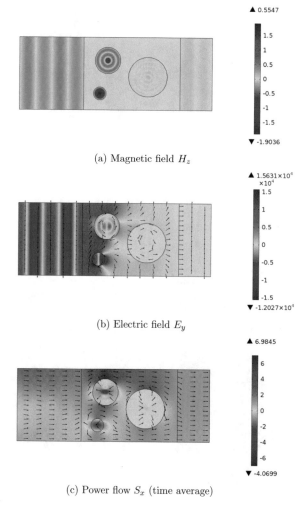

(a) Magnetic field H_z

(b) Electric field E_y

(c) Power flow S_x (time average)

3.3.1 *Electromagnetic Cloaking by Transformation Optics*

Transformation optics deals with the manipulation of the flow of light by *coordinate transformation* achieved by the employment of metamaterials. In this method, the relative permittivity and permeability of the cloak are so tailored that the incident plane wave gets bent around the object and passes undistorted. The bending is possible because the cloak distorts the space in such a manner that light is forced to avoid the object and travel around it. In a way, the region of space where the object is present becomes unavailable for the propagation of light. Since space and the associated coordinate system are being transformed, the phenomenon is called transformation optics. This method of cloaking by metamaterials was proposed by

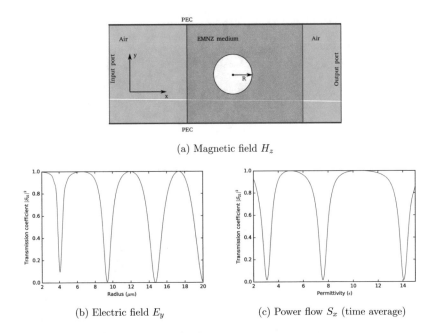

(a) Magnetic field H_z

(b) Electric field E_y

(c) Power flow S_x (time average)

Fig. 3.6 Effect of radius and permittivity of the defect in the transmission

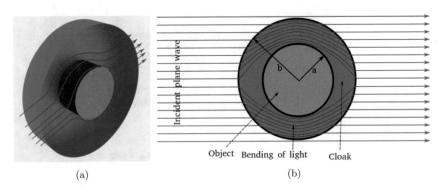

(a)

(b)

Fig. 3.7 Schematic illustration of electromagnetic cloaking by transformation optics in a **3D** and **b** 2D view

Pendry et al. in their seminal paper in the year 2006 [136]. A schematic illustration of transformation-optics-based electromagnetic cloaking is shown in Fig. 3.7, in which distortion of the path of light rays has been depicted. The object and the cloak both have cylindrical symmetry and have radii a and b, respectively.

For cloaking to occur, the components of the permittivity ($\bar{\bar{\epsilon}}$) and permeability ($\bar{\bar{\mu}}$) tensors in cylindrical coordinate system are given as [135]

$$\varepsilon_r = \mu_r = \frac{r - a}{r} \tag{3.7}$$

$$\varepsilon_\phi = \mu_\phi = \frac{r}{r - a} \tag{3.8}$$

$$\varepsilon_z = \mu_z = \left(\frac{b}{b - a}\right)^2 \frac{r - a}{r} \tag{3.9}$$

The transformation equations for Cartesian coordinate system are given as [141]

$$\varepsilon_{xx} = \varepsilon_r cos^2\phi + \varepsilon_\phi sin^2\phi \tag{3.10}$$

$$\varepsilon_{xy} = \varepsilon_{yx} = (\varepsilon_r - \varepsilon_\phi)sin\phi\, cos\phi \tag{3.11}$$

$$\varepsilon_{yy} = \varepsilon_r sin^2\phi + \varepsilon_\phi cos^2\phi \tag{3.12}$$

and

$$\bar{\bar{\varepsilon}} = \bar{\bar{\mu}} \tag{3.13}$$

Using the above equations, we numerically analyzed the phenomenon and the obtained results have been shown in Fig. 3.8. Figure 3.8a illustrates the computational cell used, in which various domains and boundaries have been highlighted. The material parameters (ϵ and μ) of the cloak were set according to Eqs. 3.7—3.12. The boundary of the embedded object was made a *perfect electric conductor* (PEC), while the top and the bottom boundaries of the computational cell were made *perfect magnetic conductor* (PMC). A z-polarized plane wave of wavelength 3μm was launched from the input port and received at the output as shown. The results of the computation are shown in Fig. 3.8b, c, as the steady-state electric field distribution throughout the cell. It can be seen that inside the cloak the wavefront is not broken but only gets deformed a bit and regains its original planar shape on emerging out of it. Hence, no light gets blocked or reflected, no shadow is cast, and the object remains absolutely cloaked. On the other hand, in the absence of the cloak, the object substantially reflects the impinging wave, the wavefront gets broken, and a shadow region gets formed. This illustrates the potential of the metamaterial of making a science fiction concept a reality.

3.3.2 Electromagnetic Cloaking by Zero Refractive Index

There is another, a newer method of achieving electromagnetic cloaking using zero-index metamaterials, popularized by Huang et al. in 2011 [27, 142–147]. They used the rods-in-air-type photonic crystal working at Dirac frequency as the zero-index metamaterial. The working principle of this type of cloaking is that the phase and magnitude of the fields remain constant throughout the zero-index medium. The phase and the magnitude of the wave at the boundary of emergence are the same

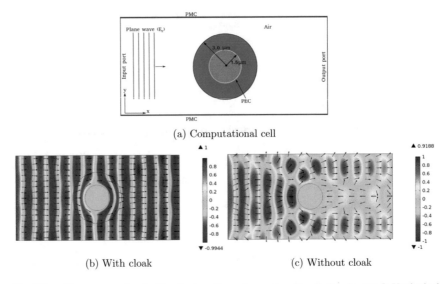

(a) Computational cell

(b) With cloak   (c) Without cloak

Fig. 3.8 **a** Electromagnetic cloaking by transformation optics. No shadow is cast. **b** Uncloaked object casts shadow. The arrows indicate the direction of power flow

as that at the boundary of incidence. This property of zero-index metamaterials has already been discussed in Chap. 2. To demonstrate the ZIM-based cloaking, we have chosen rods-in-air-type zero-index metamaterial in which an optically large object has been embedded (see Fig. 3.9a). The rods of the ZIM are made of silicon, and the surrounding medium is air. The ZIM is meant to cloak the object from any incoming radiation, thereby making it invisible. In Fig. 3.9b, the design of the computational cell used for numerical analysis has been shown, with proper boundary conditions and port locations. To thoroughly explain the cloaking effect of the ZIM, various types of simulations have been performed (Fig. 3.9c–f). Firstly, the invariance of the phase and the magnitude of the electric field throughout the zero-index metamaterial have been shown in Fig. 3.9c. Please note that the phase with which the wave enters the ZIM is the same as that with which it exits. It seems as if a part of the space has vanished for the incoming wave, by the effect of ZIM. Interestingly, the metamaterial continues to exhibit the same behavior even in the presence of an opaque object, if the ZIM region is sufficiently large, thereby cloaking it from the incident wave. Here we have used a 10 × 10 array of silicon cylinders. Figure 3.9d showcases the electromagnetic cloaking effect of the zero-index metamaterial, in which the plane wavefronts enter the ZIM and exit undistorted, without any shadow region being formed. On the other hand, in the absence of the metamaterial, the cloaking effect is absent too. Consequently, the plane wavefront gets distorted, and a shadow region gets formed behind the object (Fig. 3.9e). This illustrates the effectiveness of the ZIM in providing the cloaking effect. Finally, for the sake of comparison and confirmation, we performed the computation using a homogeneous ZIM slab of the same size as the previously used 10 × 10 array, perfect exhibition of cloaking (Fig. 3.9f) with

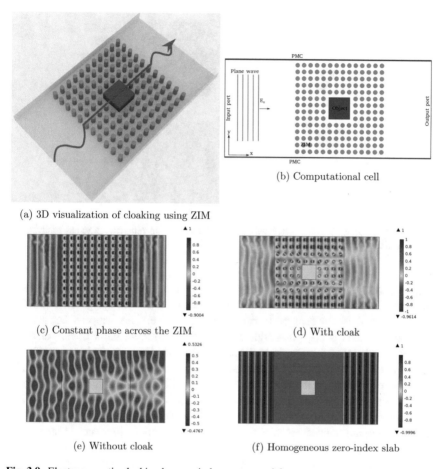

(a) 3D visualization of cloaking using ZIM

(b) Computational cell

(c) Constant phase across the ZIM

(d) With cloak

(e) Without cloak

(f) Homogeneous zero-index slab

Fig. 3.9 Electromagnetic cloaking by zero-index metamaterial

perfectly shaped planar wavefronts is observed. All these results and discussion establish the electromagnetic cloaking capability of zero-index metamaterials, which enhances their attractiveness.

3.4 Wavefront Engineering

In general, the shape of a wavefront is governed by two key aspects—the type of the source and the shape of the object the wave interacts with. A point source generates a spherical wavefront, a line source generates a cylindrical wavefront, and any type of wavefront transforms into planar on traveling sufficiently large distance away from the source [90, 148–151]. A common and conventional method of altering the

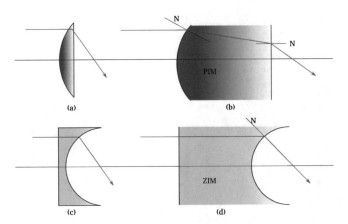

Fig. 3.10 Schematic illustration of how a ZIM slab having a specific curvature can equivalently behave as a lens

shape of the wavefront is by using lenses. A convex lens can convert a diverging wavefront into a planar one and a plane wavefront to a converging one depending on the position of the lens with respect to the source. This happens because the shape of the lens is such that light falling on its different parts travels different optical path lengths through it, and hence refract at different angles according to Snell's law [14]. Recently, it has been shown that the shape of the wavefront can be manipulated by making the wave pass through the ZIM slab of a particular shape, or more precisely a *ZIM lens*. The technique is called *wavefront engineering*, and it is emerging as a potential application of zero-index metamaterials. The figures below and the following discussion will present a clear picture of the mechanism behind the wavefront manipulation using both the conventional lenses and the ZIM lenses. Figure 3.10 shows the ray optics picture of the convergent nature of conventional convex lens and a concave zero-index metamaterial lens. Despite their different geometry, both the lenses create the same effect. Figure 3.10a, c shows the outside picture of the bending of light rays on passing through the two types of lenses, while Fig. 3.10b, d shows the in-depth mechanism of refraction by them according to Snell's law. The convex lens shown in the left converge the incident parallel ray as usual. But the ZIM lens on right also exhibits the converging effect, despite having a concave surface, which opposite to what is expected from a conventional lens of the same shape made of a positive-index material. In other words, a conventional plano-concave lens is diverging, whereas a plano-concave ZIM lens is converging. The converging nature of the shown ZIM lens is attributed to the property of the normal emergence of light from a zero-index medium, as discussed in Chap. 2.

Figure 3.10 presents only a schematic picture of the behavior of the ZIM lens, predicted by Snell's law. In addition to this, we can also perform a full-wave analysis to investigate how ZIM lenses can be employed to reshape the wavefront at will. We performed the numerical analysis, whose results have been shown below. In Fig. 3.11a, a pano-concave ZIM lens has been used and a plane wave has been made

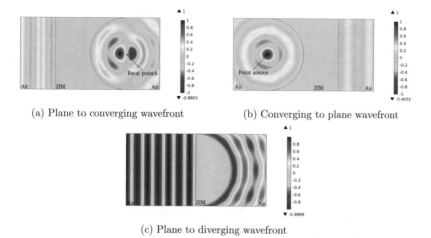

(a) Plane to converging wavefront

(b) Converging to plane wavefront

(c) Plane to diverging wavefront

Fig. 3.11 Reshaping the wavefront using different types of zero-index metamaterial lenses

to impinge on its flat surface. We observe that light emerges from the concave side as a train of cylindrical wavefronts and converges to a focal point (as illustrated). In this way, the ZIM lens transform the plane wavefront into a cylindrical wavefront. Figure 3.11b shows the reverse operation, where spherical wavefronts originating from a point source hit the concave side, pass through the zero-index medium, and emerge as plane wavefronts from the flat side. This verifies the reversibility, also shown by the conventional lenses. Figure 3.11c shows a different type of lens—a plano-convex lens—which has been shown to convert plane wavefronts hitting on the plane side to cylindrical wavefronts emerging from the convex side. One can infer from the above results that a concave ZIM lens is converging while a convex ZIM lens is diverging, which is absolutely opposite to the case of conventional lenses.

3.5 Directive Radiation by Manipulation of Radiation Pattern

As explained in Chap. 2, inside a ZIM, a wave experiences zero change in phase, i.e., all the points on the boundary and in the bulk of the ZIM vibrate in the same phase. This property can be exploited to generate highly directive radiation [25, 65, 152, 153]. If one embeds a point source inside the ZIM slab, the whole slab vibrates in phase with the point source, behaving like an extended source and emitting radiation in all directions, albeit unequally. The amount of radiation emerging from a boundary of the ZIM slab depends on its length. A longer boundary has a greater number of points acting as individual point sources, and hence emits greater power than a shorter one.

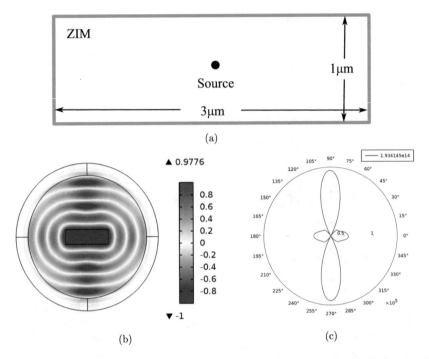

Fig. 3.12 Demonstration of the capability of a homogeneous ZIM slab to generate highly directional beams of light. **a** Geometry of the slab with an embedded current line source, **b** Electric field distribution around it, and **c** far-field plot to display the amount of power flow w.r.t. the azimuthal coordinate at 193.4 THz (1550 nm)

To numerically demonstrate the capability of directive radiation, we choose a rectangular slab of aspect ratio 3 : 1 of a homogeneous zero-index medium, and insert a current line source at its geometrical center. The design of the computation cell has been shown in Fig. 3.12a. The current line source acts as a point source and makes the whole slab vibrate in coherence with it. Figure 3.12b, c illustrates the field around the slab and the radiation pattern, respectively. It can be observed in Fig. 3.12c that a significantly greater amount of radiation emerges out of the longer edge of the ZIM slab compared to the shorter one. In light of the above explanation, it seems logical that the longer boundary, having more number of points, gives out a greater amount of radiation compared to the shorter one. In this way, we achieved a technique of controlling the radiation pattern of a point source like a quantum emitter, and obtain highly directive radiation from an otherwise almost omnidirectional source.

3.6 Beam Splitting and Beam Steering

Based on the zero-phase change property of a zero-index medium, two more use-ful applications can be *beam splitting* and *beam steering* [154–162]. Beam splitting means dividing an incident beam into multiple parts, while beam steering means con-trolling the orientation of a beam and hence the direction propagation of power. The zero-phase change and the normal emergence properties of a zero-index metamate-rial facilitate both applications [163, 164]. We know that each face (or boundary) of a zero-index slab acts as an independent source. Light entering through one face emerges from all the remaining faces. Hence, by controlling the number of faces of a ZIM slab, the number of emergent beams can be controlled, and by controlling the area of the faces, the power of the beam can be controlled. If all output faces are of equal area, multiple beams of equal power emerge from the ZIM slab. In other words, the power of the input beam gets distributed equally among all the output beam. We utilized this idea to create ZIM-based beam splitters, whose design schematics and numerical results are shown in Fig. 3.13. We simulated two types of beam splitters—a one-to-two splitter and a one-to-three splitter. The former splits the input beam into two equal parts while the latter divides it into three equal parts.

Figure 3.13a shows the schematic of one-to-two splitter. Light propagating in a silicon waveguide is fed into a zero-index right-angled isosceles prism from the hypotenuse side and emerged from the two equal sides. The output radiation is collected and measured at port 2 and port 3. Figure 3.13b shows the schematic of the one-to-three splitter, in which light was fed from port 1, propagated toward the ZIM cube, and emerged from the remaining three sides as shown. The emerged beams were collected and measured at the three output ports, i.e., port 2, port 3, and port 4. The output power at the output ports has been mentioned in Table 3.1 for both the types of beam splitter. In both the designs, the output beams have equal distribution of power, which is favorable for on-chip employment in photonic integration. In the case of the one-to-three splitter, the power at port 3 is minutely greater than that at port 2 and port 4 because of a minute bending loss for the latter two ports.

Next, we demonstrate the phenomenon of beam steering based on the same two properties of zero phase change and normal emergence. By employing a zero-index medium, there can be two routes of achieving the beam-steering capability. One is by using a zero-index prism and manipulating its geometry, and the other is by using a zero-index slab and varying its permittivity. Both the schemes have been shown in Fig. 3.14. The first method involves the use of a zero-index right-angled prism, whose base angle A is varied from 0° to 45°. The variation of A results in the change in the slope of the hypotenuse, which further results in the change in the orientation of the normally emerging beam. The input, in this case, is fed from the bottom boundary of the ZIM prism. The second method involves a slab, whose refractive index changes from 0.001 to 1.0. The variation of the refractive index changes the angle of refraction of the emergent beam. The input, in the second case, is fed from the left-hand side boundary of the ZIM. In both the designs, the regions above and below the ZIM are air, and the upper air regions have been backed by

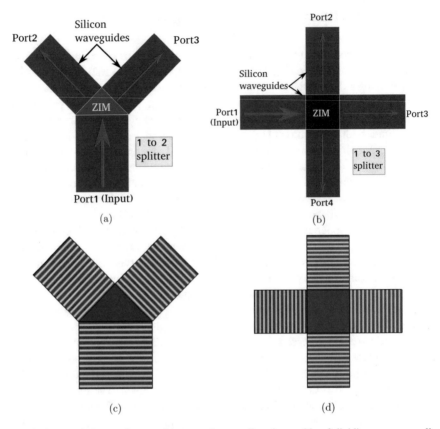

Fig. 3.13 Zero-index medium working as a beam splitter is capable of dividing power equally among all the output beams

Table 3.1 The distribution of power at different output ports with respect to the input port for two types of beam splitters

	Port 1 (input)	Port 2	Port 3	Port 4
One-to-two splitter	P_0	$0.3181P_0$	$0.3181P_0$	–
One-to-three splitter	P_0	$0.16477P_0$	$0.16551P_0$	$0.16477P_0$

perfectly matched layer (PML) [44, 165]. We numerically analyzed both the routes and have demonstrated the techniques below in Figs. 3.15 and 3.16.

Figure 3.15a–d shows the distribution of the electric field in the far-field region, illustrating the orientation of the beam. For $A = 0$, the prism becomes a horizontal slab, and hence the beam is vertical. As the angle A increases to 15, 30, 45 degrees, the orientation of the means shifts away from the vertical orientation and approaches

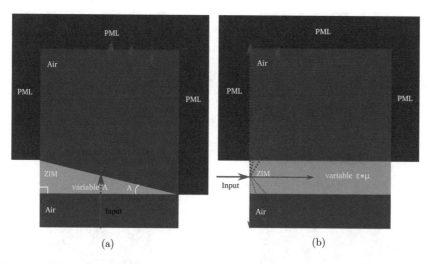

Fig. 3.14 The schematic illustration of two types of beam steering. **a** By varying the angle A of the ZIM prism and **b** by varying the refractive index $n = \sqrt{\epsilon\mu}$ of the slab, keeping $\epsilon = \mu$

the horizontal direction. The far-field polar plot of Fig. 3.15e shows the reduction in the angle of elevation of the beam from 90° to 45°, as the angle A increases from 0° to 45°. In this way, beam steering is achieved by varying the slope of the emergence boundary.

In the second technique, the permittivity ϵ and the permeability μ are simultaneously varied from 0.001 to 1.0, such that $\epsilon = \mu$. Therefore, the refractive index $n = \sqrt{\epsilon\mu}$ also gets varied through the same values. In Fig. 3.16a–e, the refractive index acquires the values—0.001, 0.01, 0.1, 0.5, and 1.0, respectively. In Fig. 3.16a, when $n = 0.001 \approx 0$, the beam is vertical. As the refractive index increases, the beam starts shifting away from the vertical orientation and begins to approach the horizontal orientation. In Fig. 3.16a–d, intense beams of light form and are oriented in particular directions, but in Fig. 3.16e, light is spread in almost all the directions without the formation of a single well-defined beam. It happens because, in the last case, the refractive index of the slab is 1.0, i.e., the same as that of air. Hence, light sees the entire computational region as a single medium, and *no refraction takes place*. Whereas in all the other cases, the refractive index of the slab is different from the surroundings, the refraction does take place. Therefore, in Fig. (a)–(d), well-defined beams are formed, and in Fig. (e) light spreads in all the directions. The same can be observed in the polar plot of Fig. 3.16f too. In this way, beam steering can be achieved by manipulating the refractive index of the slab to values less than 1. Beam splitting and beam steering are important applications from the integrated photonics point of view.

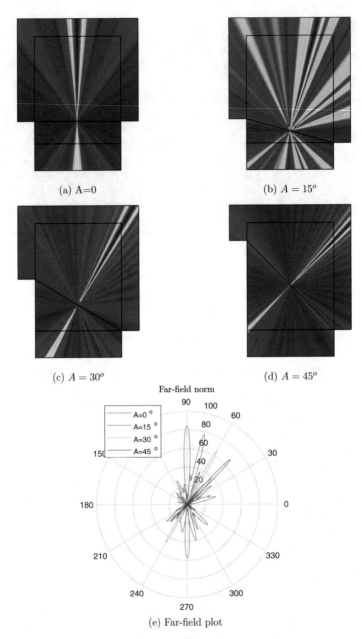

(a) A=0   (b) $A = 15^o$

(c) $A = 30^o$ (d) $A = 45^o$

(e) Far-field plot

Fig. 3.15 The demonstration of beam steering using epsilon-mu-near-zero metamaterials by varying the angle of prism from $0°$ to $45°$

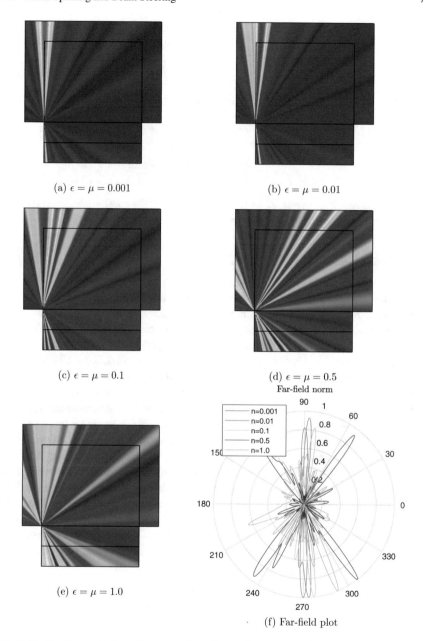

(a) $\epsilon = \mu = 0.001$

(b) $\epsilon = \mu = 0.01$

(c) $\epsilon = \mu = 0.1$

(d) $\epsilon = \mu = 0.5$

(e) $\epsilon = \mu = 1.0$

(f) Far-field plot

Fig. 3.16 The demonstration of beam steering using epsilon-mu-near-zero metamaterials by varying the effective index from 0.001 to 1.0

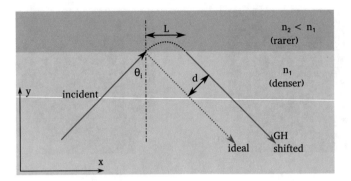

Fig. 3.17 Schematic illustration of normal Goos–Hänchen shift. The actually reflected ray is shifted from the ideal case by a perpendicular distance $'d'$ and a lateral distance $'L'$

3.7 Zero Goos–Hänchen Shift

According to the ray optics perspective, if light is traveling from a denser medium to a rarer medium and the angle of incidence is greater than the critical angle, all the light is reflected back into the denser medium, and the phenomenon is called the *total internal reflection*. Ideally, no light should be able to penetrate into the rarer medium, and the light should get reflected exactly from the point of incidence. However, in reality, some light does penetrate into the rarer medium, travels some distance along the interface, and then returns to the denser medium. Consequently, the point of reflection is a bit shifted from the point of incidence, and the phenomenon is called the *Goos–Hänchen shift* [166, 167]. The amount of shift depends on the refractive indices of the two media, the angle of incidence, and the polarization of light.

Above was the case of Fig. 3.17 which shows the schematic illustration of the Goos–Hänchen shift. It shows an incident ray hitting the interface, penetrating into the rarer medium to some extent, and getting reflecting back into the denser medium with a certain lateral shift. It also shows the ideal case that would have been had the ray not penetrated into the second medium. The distance between the ideally reflected ray and the actually reflected ray is the Goos–Hänchen shift and has been denoted by d. According to Ghatak and Thyagarajan (1978) [168], Goos–Hänchen shift is given for *p-polarization* as

$$d_p = \frac{\lambda_1}{\pi}(n^2 - 1)\frac{tan\,\theta_i}{(sin^2\,\theta_i - sin^2\,\theta_c)^{1/2}[cos^2\,\theta_i + n^4(sin^2\,\theta_i - sin^2\,\theta_c)]} \quad (3.14)$$

and for *s-polarization*[1] as

$$d_s = \frac{\lambda_1}{\pi}\frac{tan\,\theta_i}{(sin^2\,\theta_i - sin^2\,\theta_c)^{1/2}} \quad (3.15)$$

---

[1] In p-polarization, **E** is parallel to the plane of incidence, while in s-polarization it is perpendicular.

where $\lambda_1 = \lambda_0/n_1$ is the wavelength in medium 1, n_1 is the refractive index of medium 1, n_2 is the refractive index of medium 2, $n = n_1/n_2$, θ_i is the angle of incidence, and $\theta_c = sin^{-1}(n_2/n_1)$ is the critical angle. This is the microscopic view of total internal reflection with natural materials. The phenomenon takes an interesting modification when the second medium is replaced by an epsilon-near-zero (ENZ) medium.

3.7.1 Inside the Epsilon-Near-Zero Medium

According to Snell's law, the critical $\theta_c = sin^{-1}(n_2/n_1) = sin^{-1}(\epsilon_2/\epsilon_1)$, if $\mu_2 = \mu_1 = \mu = 1$. When the second medium is of near-zero permittivity, i.e., $\epsilon_2 \approx 0$, then $\theta_c \approx 0$. This means that at any angle of incidence except zero, the light will be total internally reflected when it travels from a positive-epsilon medium (glass in this case) to an ENZ medium. For such an arrangement, Xu et al. expressed the equations of Goos–Hänchen shift in terms of epsilon [169] as

$$d_s = \frac{\lambda}{\pi} \frac{sin\,\theta_i}{\sqrt{sin\,\theta_i - \epsilon_2}} \tag{3.16}$$

for s-polarization and

$$d_p = \frac{\lambda}{\pi} \frac{\epsilon_2(1 - \epsilon_2)sin\,\theta_i}{(\epsilon_2^2 cos^2\,\theta_i + sin^2\,\theta_i - \epsilon_2)\sqrt{sin^2\,\theta_i - \epsilon_2}} \tag{3.17}$$

for p-polarization. According to the above formulae, if the permittivity of the second medium vanishes, i.e., $\epsilon_2 \to 0$, so does the p-polarization GH shift, i.e., $d_p \to 0$, but there is still a finite shift ($\approx \lambda/\pi$) experienced by s-polarization, independent of the angle of incidence. Figure 3.18 schematically shows how differently the two

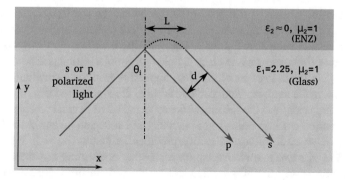

Fig. 3.18 Schematic illustration how Goos–Hänchen shift, with an ENZ medium as cladding for s- and p-polarizations

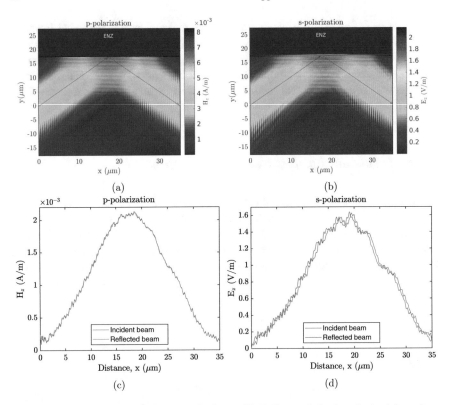

Fig. 3.19 Distribution of **a** H_z for p-polarization and **b** E_z for s-polarization obtained throughout the two media during total internal reflection. Field distribution corresponding to incident and reflected beams along x-direction for **c** p-polarization and **d** s-polarization

polarizations reflect from the interface. Medium 1 is glass with permittivity $\epsilon_1 = 2.25$ and permeability $\mu_1 = 1$ (being non-magnetic), and medium 2 is an epsilon-near-zero medium with permittivity $\epsilon_2 \approx 0$ and permeability $\mu_2 = 1$. The p-polarized wave gets reflected ideally, without any shift, whereas the s-polarized one does have a finite GH shift. The arrangement shown in Fig. 3.18 on being analyzed numerically yields the results shown in Fig. 3.19. Figure 3.19a, b shows the distribution of z-component of the magnetic and electric fields for p-polarization and s-polarization, respectively. In the case of p-polarization, no field is seen to be penetrating into the ENZ region, while there is a noticeable leaking of the s-polarized field into the ENZ cladding which results in the shifting of the reflected wave. A closer inspection of the field distribution corresponding to the incident and reflected beams at the interface between the two media along x-direction provides the graphs shown in Fig. 3.19c–d. In Fig. 3.19c, the two graphs overlap on each other confirming the absence of any shift, whereas in Fig. 3.19d there is a noticeable shift in the reflected beam with respect to the incident beam at the interface. By precise measurement, the lateral shift was found to be $L = 0.4$ μm, and the corresponding Goos–Hänchen shift is determined to be $d_s = L \cos \theta_i \approx 0.2828$ μm. The free-space wavelength λ_0 used in this simulation is 1 μm, which reduced to $\lambda_1 = \lambda_0/1.5$ inside medium 1. The angle of incidence $\theta_i = 45°$, the permittivity $\epsilon_2 = 0.001$, and scattering boundary condition have been used on all the four sides.

The leakage of power into the cladding results in a reduction of the intensity of a reflected beam, which becomes a major drawback in systems involving multiple reflections of light, where it propagates while losing some power (however small it is) on every reflection. The use of zero-index metamaterials provides an efficient solution for this problem.

3.8 Thresholdless Lasers

We know that in a crystal lattice, there are several energy levels available for atoms to occupy. Atoms can absorb energy from outside and excite to higher energy levels or release energy and de-excite to lower energy levels. According to Boltzmann's law, the number of atoms occupying higher energy levels is relatively lesser than those occupying lower energy levels, at thermal equilibrium [170–172]. Figure 3.20 shows two energy levels, E_1 and E_2, having population densities N_1 and N_2, respectively. According to Boltzmann's law, $N_2/N_1 = e^{-(E_2-E_1)/K_BT}$, hence $N_2 < N_1$. Between the two shown energy levels, there are three types of transitions possible—*absorption*, *spontaneous emission*, and *stimulated emission*. Atoms of energy level E_1 can absorb incoming radiation of frequency $\omega = (E_1 - E_2)/\hbar$ and excite to the E_2. On the other hand, an atom can de-excite from level 2 to level 1 by releasing energy $E_1 - E_2$, or in other words, a photon of frequency $\omega = (E_1 - E_2)/\hbar$. While absorption can only be of the stimulated type, emission can be either spontaneous or stimulated. The spontaneous emission happens on its own, depends on the number of atoms in the excited state only and does not require the presence of the optical field. Whereas the stimulated emission requires radiation of a particular frequency to occur, and hence depends on the relative power of various frequencies of the spectrum as well as on the number of atoms in the excited state [11, 173]. In Optical Electronics (2011), Ghatak and Thyagarajan have presented a detailed analysis of atomic transitions inside a laser system, using Einstein's A and B coefficients [11]. In a system illustrated in Fig. 3.20, the number of stimulated transitions $1 \rightarrow 2$ is given by

$$\Gamma_{12} = B_{12} N_1 u(\omega) \tag{3.18}$$

the number of stimulated transitions $2 \rightarrow 1$ is given by

$$\Gamma_{21} = B_{21} N_2 u(\omega) \tag{3.19}$$

and the number of spontaneous transitions $2 \rightarrow 1$ is given by

$$U_{21} = A_{21} N_2 \tag{3.20}$$

At thermal equilibrium, the total number of transitions from 1 to 2 is equal to the total number of transitions from 2 to 1, i.e.,

$$\Gamma_{12} = \Gamma_{21} + U_{21} \tag{3.21}$$

$$B_{12} N_1 u(\omega) = B_{21} N_2 u(\omega) + A_{21} N_2 \tag{3.22}$$

or

$$u(\omega) = \frac{A_{21}}{(N_1/N_2) B_{12} - B_{21}} \tag{3.23}$$

$$= \frac{A_{21}}{B_{12} e^{\hbar\omega/K_B T} - B_{21}} \tag{3.24}$$

According to Planck's law of radiation, the energy density per unit frequency interval is given as [174, 175]

$$u(\omega) = \frac{\hbar\omega^3 n_0^3}{\pi^2 c^3} \frac{1}{e^{\hbar\omega/K_B T} - 1} \tag{3.25}$$

where $K_B = 1.38 \times 10^{-23} J/K$ is Boltzmann's constant, c is the velocity of light, and n_0 is the refractive index of the medium. Comparing Eq. 3.24 and Eq. 3.25, we get

$$B_{12} = B_{21} = B \tag{3.26}$$

and

$$\frac{A_{21}}{B_{21}} = \frac{A}{B} = \frac{\hbar\omega^3 n_0^3}{\pi^2 c^3} \tag{3.27}$$

Equation 3.26 tells us that the rate of stimulated absorption is equal to that of the stimulated emission, and Eq. 3.27 presents the ratio of spontaneous to stimulated emission rates. At thermal equilibrium, the ratio of the number of spontaneous emission to the number of stimulated emission is given as

$$R = A_{21} N_2 / B_{21} N_2 u(\omega) = e^{\hbar\omega/K_B T} - 1 \tag{3.28}$$

And it can be shown that (see Ghatak and Thyagarajan (2011) [11]) in an optical source at temperature $T = 1000$ K, emitting radiation of wavelength $\lambda \approx 500$ nm, the ratio $R \approx 5.0 \times 10^{12}$. It means that the emission from the mentioned optical source is predominantly spontaneous, hence incoherent. Here comes an interesting vision for such a system! From Eq. 3.27, we can notice that A/B is directly proportional to n_0^3. If n_0 is reduced by a factor of 10, the A/B ratio is reduced drastically by a factor

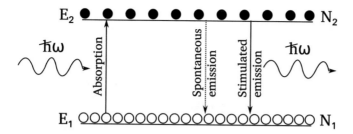

Fig. 3.20 Possible transitions of atoms between two energy levels on interaction with light of frequency corresponding to the energy difference between two levels

of 1000. It means that if the refractive index of the medium of the optical source vanishes, the spontaneous emission can get suppressed in favor of the stimulated emission, and the lasing action is easily achievable. This assertion gets concretized if one analyzes the situation in terms of population inversion.

Let us consider a laser cavity consisting of two mirrors enclosing a slab of an active medium between them, as shown in Fig. 3.21. The length of the cavity is d, the reflectivity of the primary mirror is $R_1 = 1$, and that of the secondary mirror (also called *output coupler*) is $R_2 \approx 1$. The role of the active medium is to amplify the optical signal oscillating between the mirrors. Let this be a three-level laser system and the set of energy levels shown in Fig. 3.20 be one involved in the lasing action. When sufficiently high population inversion is achieved, i.e., $N_2 > N_1$ and the population of level 2 increases beyond a certain threshold, i.e., $N_2 - N_1 > \Delta N_{th}$, the system can amplify any signal of the frequency corresponding to the energy bandgap, because the presence of the optical signal stimulates all the atoms present in the level E_2 to de-excite simultaneously to level E_1, thereby producing a highly intense, coherent, and parallel beam of light of the same frequency. The process is thus called stimulated emission and the device is aptly referred to as *LASER (light amplification by stimulated emission of radiation)* [176–179]. For broad details of laser mechanism read ref [11]. An important parameter to be considered here is the threshold population inversion ΔN_{th}, below which the optical signal succumbs to various losses inside the cavity. It can be shown that [11] the threshold population inversion is given by

$$\Delta N_{th} = \frac{\omega^2 n_0^3}{\pi^2 c^3} \frac{t_{sp}}{t_c} \frac{1}{g(\omega)} \tag{3.29}$$

where ω is the frequency of light, c is the speed of light, n_0 is the refractive index of the active medium, $t_s p$ is the spontaneous lifetime, i.e., the time for which an electron stays in the excited state before undergoing a spontaneous transition to ground state accompanied by emission, t_c is the cavity lifetime, i.e., the time in which the energy in the cavity is reduced by a factor of e^{-1} and $g(\omega) = 1/\Delta\omega$ is the line-shape function, $\Delta\omega$ being the absorption bandwidth of the active medium around ω.

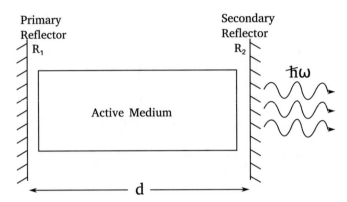

Fig. 3.21 Construction of a basic optical resonator consisting of two mirrors enclosing a slab of an active medium between them

In modern times of low power devices, it is desirable that ΔN_{th} should be as low as possible. The lower the ΔN_{th}, the lesser the power required to achieve it! For example, a Ruby laser has ΔN_{th} of the order of 10^{17} cm^{-3}, while that of a He-Ne laser is of the order of 10^8 cm^{-3}. Clearly, a He-Ne [180] laser works at a lower power supply than a Ruby laser [178], and hence is a preferred choice. ΔN significantly depends on $\Delta \omega$, t_c, and $t_s p$, and hence by using a cavity with a longer cavity lifetime (t_c) and an active medium with a smaller relaxation time (t_{sp}) and a narrower absorption bandwidth ($\Delta \omega$), the threshold value can be reduced significantly. Another important parameter which could not be manipulated earlier but is now controllable by using metamaterial is the refractive index. Refractive index can play a major role in deciding the value of threshold, since ΔN_{th} is directly proportional to cube of n_0, which means that by reducing n_0 by a factor of 10, ΔN_{th} can be reduced by a factor of 1000, as was seen in case of A/B ratio. In fact, if refractive index is diminished down to almost zero, ΔN_{th} will also be pushed to negligibly small value. In other words, an active medium with effectively zero refractive index will have no threshold value of population inversion to begin lasing action, which means the advent of *thresholdless lasers* working on extremely low powers [181, 182]. Such laser sources can prove revolutionary in the near future, which is expected to be full of nanotechnology and nanophotonic devices. The potential of zero-index metamaterials in laser cavities is promising and if implemented successfully will be revolutionary, and hence is worth investigating.

Chapter 4
Nonlinear Optics with Zero-Index Metamaterials

4.1 What Is Nonlinear Optics?

When any material is subject to an external electric field, electrons of each atom get displaced by a certain amount depending on the strength of the electric field. Displacement of the electron cloud results in the separation of positive and negative charge centers, thus inducing a polarization.

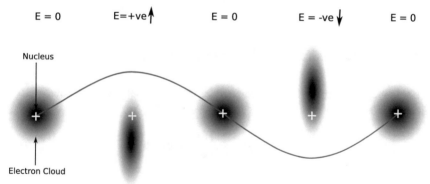

Polarization $\tilde{P}$ is written as a function of the applied electric field as [1]

$$\tilde{P}(t) = \epsilon_0(\chi^{(1)} \tilde{E}(t) + \chi^{(2)} \tilde{E}^2(t) + \chi^{(3)} \tilde{E}^3(t) + \cdots) \tag{4.1}$$

in which the first term represents the linear dependence of polarization on the electric field and the second term onward represents the nonlinear behavior. In general, $\chi^{(n)}$ is the nth-order nonlinear susceptibility. Hence, Eq. 4.1 can be written as

$$\tilde{P}(t) = \tilde{P}^{(1)} + \tilde{P}^{NL} \tag{4.2}$$

---

[1] Tilde sign on P, E, and D indicates time dependence.

where $\tilde{P}^{(1)} = \epsilon_0 \chi^{(1)} \tilde{E}(t)$ is the linear polarization and $\tilde{P}^{NL}$ is the nonlinear polarization due to the rest of the terms. The net displacement field $\tilde{D}$ is written as

$$\tilde{D}(t) = \epsilon_0 \tilde{E}(t) + \tilde{P}(t) \tag{4.3}$$

Normally, for relatively low electric fields, only the linear part $\tilde{P}^{(1)}$ is significant and the nonlinear part $\tilde{P}^{NL}$ can be ignored due to small values. Then,

$$\tilde{D}(t) = \tilde{D}^{(1)}(t) \tag{4.4}$$

$$= \epsilon_0 \tilde{E}(t) + \tilde{P}^{(1)}(t) \tag{4.5}$$

$$= \epsilon_0 \tilde{E}(t) + \epsilon_0 \chi^{(1)} \tilde{E}(t) \tag{4.6}$$

$$= \epsilon_0 (1 + \chi^{(1)}) \tilde{E}(t) \tag{4.7}$$

$$= \epsilon_0 \epsilon^{(1)} \tilde{E}(t) \tag{4.8}$$

where $\epsilon^{(1)}$ is the first-order relative permittivity. However, in the case of intense electric fields, as in a laser, the nonlinear part begins to play a significant role and gives rise to several interesting phenomena. This chapter is dedicated to the study of those nonlinear phenomena and how useful the zero-index metamaterials can be for them.

4.1.1 Wave Equation in a Nonlinear Medium

From the various texts on electromagnetics, we know that the wave equation in a medium is written as

$$\nabla^2 \tilde{\mathbf{E}} - \mu \epsilon \frac{\partial^2 \tilde{\mathbf{E}}}{\partial t^2} = 0 \tag{4.9}$$

or

$$\nabla^2 \tilde{\mathbf{E}} - \frac{\mu_r \epsilon_r}{c^2} \frac{\partial^2 \tilde{\mathbf{E}}}{\partial t^2} = 0 \tag{4.10}$$

For a non-magnetic medium, i.e., $\mu_r = 1$,

$$\nabla^2 \tilde{\mathbf{E}} - \frac{\epsilon_r}{c^2} \frac{\partial^2 \tilde{\mathbf{E}}}{\partial t^2} = 0 \tag{4.11}$$

Multiplying and dividing the second term by ϵ_0,

$$\nabla^2 \tilde{\mathbf{E}} - \frac{\epsilon}{\epsilon_0 c^2} \frac{\partial^2 \tilde{\mathbf{E}}}{\partial t^2} = 0 \tag{4.12}$$

Absorbing the ϵ into the differential, we get

$$\nabla^2 \tilde{\mathbf{E}} - \frac{1}{\epsilon_0 c^2} \frac{\partial^2 \tilde{\mathbf{D}}}{\partial t^2} = 0 \tag{4.13}$$

Now, since $\tilde{\mathbf{D}} = \epsilon_0 \tilde{\mathbf{E}} + \tilde{\mathbf{P}}^{(1)} + \tilde{\mathbf{P}}^{NL}$ (from Eqs. 1.2, 1.3, 1.4, and 1.8), wave equation becomes

$$\nabla^2 \tilde{\mathbf{E}} - \frac{1}{\epsilon_0 c^2} \frac{\partial^2 \tilde{\mathbf{D}}^{(1)}}{\partial t^2} = \frac{1}{\epsilon_0 c^2} \frac{\partial^2 \tilde{\mathbf{P}}^{NL}}{\partial t^2} \tag{4.14}$$

or

$$\nabla^2 \tilde{\mathbf{E}} - \frac{\epsilon^{(1)}}{c^2} \frac{\partial^2 \tilde{\mathbf{E}}}{\partial t^2} = \frac{1}{\epsilon_0 c^2} \frac{\partial^2 \tilde{\mathbf{P}}^{NL}}{\partial t^2} \tag{4.15}$$

The wave equation obtained as Eq. 4.15 is different from the usual form, as it has a source term on the right-hand side, making it an inhomogeneous differential equation. This form of the wave equation is used in the mathematical analysis of all nonlinear phenomena.

4.2 Nonlinear Phenomena

Certain materials have considerable values of second-order and third-order susceptibilities. Such materials, when subject to a sufficiently intense beam of light, exhibit interesting phenomena like second-harmonic generation, sum-frequency generation, difference-frequency generation, four-wave mixing, third harmonic, optical bi-stability, etc. Below, we have explained these phenomena so that when these concepts are invoked in association with zero-index metamaterials in the later parts of this chapter, the reader does not find them difficult to grasp. For the rigorous analysis of the principles on nonlinear optics, it is advisable to read Boyd [183].

4.2.1 Second-Harmonic Generation

When some amount of light of a certain frequency ω is converted to that of frequency 2ω, owing to the considerable contribution of the second-order susceptibility, it is called *second-harmonic generation* [184]. Here, ω is the fundamental frequency and

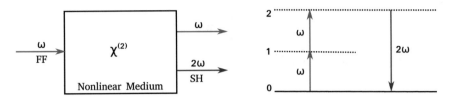

Fig. 4.1 Second-harmonic generation: **a** Schematic illustration **b** Energy band diagram. FF = fundamental frequency, SH = second harmonic

2ω is the second-harmonic frequency. Figure 4.1 illustrates the process of second-harmonic generation by means of a block diagram and an energy band diagram. The figure shows that when the fundamental frequency (ω) is fed as input into a nonlinear medium of high second-order susceptibility, a fraction of it gets converted into second harmonic (2ω) and the output is a mixture of two frequencies ω and 2ω. The band diagram shows the mechanism of the conversion process, in which it can be seen that two photons of frequency ω are absorbed and an atom is excited to energy level 2, and then the atom de-excites to the ground level, and one photon of frequency 2ω is emitted. The percentage of the input power converted into the second harmonic is called the *conversion efficiency* of the process. The mathematical analysis of the second-harmonic generation process is given as follows.

Let the input electric field of the fundamental frequency ω be given as

$$\tilde{E}(t) = Ee^{-i\omega t} + c.c. \tag{4.16}$$

where $c.c.$ is the complex conjugate of the preceding quantity [11, 183]. The contribution of the second-order susceptibility into the polarization is

$$\tilde{P}^{(2)}(t) = \epsilon_0 \chi^{(2)} [\tilde{E}(t)]^2 \tag{4.17}$$

$$= \epsilon_0 \chi^{(2)} [Ee^{-i\omega t} + c.c.]^2 \tag{4.18}$$

$$= \epsilon_0 \chi^{(2)} [Ee^{-i\omega t} + E^* e^{i\omega t}]^2 \tag{4.19}$$

$$= \epsilon_0 \chi^{(2)} [E^2 e^{-2i\omega t} + E^{*2} e^{2i\omega t} + 2EE^*] \tag{4.20}$$

$$= \epsilon_0 \chi^{(2)} [2EE^* + E^2 e^{-2i\omega t} + E^{2*} e^{2i\omega t}] \tag{4.21}$$

$$= \epsilon_0 \chi^{(2)} 2EE^* + [\epsilon_0 \chi^{(2)} E^2 e^{-2i\omega t} + c.c.] \tag{4.22}$$

The first term on the right-hand side of Eq. 4.22 is the zero-frequency term and of not much relevance here, while the second term represents the existence of the second-harmonic frequency. A common practical example of second-harmonic generation is the production of 530 nm green laser from Nd:YAG [185–187] laser of infrared wavelength 1060 nm. The second-harmonic generation has been a very useful technique to develop lasers of desired frequencies.

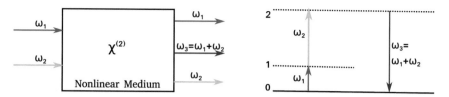

Fig. 4.2 Sum-frequency generation: **a** Schematic illustration **b** Energy band diagram

4.2.2 Sum- and Difference-Frequency Generation

4.2.2.1 Sum-Frequency Generation

Sum-frequency generation (SFG) is similar to second-harmonic generation, the only difference is that in place of a single input frequency, now there are two input frequencies ω_1 and ω_2. Figure 4.2 schematically illustrates the process of sum-frequency generation with the help of a block diagram and an energy band diagram. It can be seen that two input frequencies are fed into the nonlinear system of significant second-order susceptibility and a mixture of three frequencies is obtained on the output side. The three components of the output are the two original frequencies ω_1 and ω_2, and a third-frequency ω_3 which is the sum of the two input frequencies, i.e., $\omega_3 = \omega_1 + \omega_2$. Due to this reason, the process is also called *three-wave mixing* [188, 189]. The energy band diagram in Fig. 4.2b shows the mechanism behind the sum-frequency generation. Absorption of a photon of frequency ω_1 excites the atom from the ground state to energy level 1, which, if followed by absorption of a photon of frequency ω_2, further excites the atom to energy level 2. The excited atom returns to the ground state releasing a photon of the sum-frequency $\omega_3 = \omega_1 + \omega_2$, which is greater than either of the two input frequencies. It is important to understand here that the second-harmonic generation can be considered as a special case of sum-frequency generation, where the two input frequencies are equal. Sum-frequency generation is a useful technique of obtaining a tunable high-frequency laser, if one of the inputs is a fixed low-frequency laser whereas the other one is fed by a tunable low-frequency laser.

4.2.2.2 Difference-Frequency Generation

Difference-frequency generation (DFG), as the name suggests, is the reverse of sum-frequency generation. In DFG, the input comprises a high frequency and a low frequency, and the output contains the two input frequencies and a third frequency which is the difference of the two input frequencies. In this way, DFG is a *three-wave mixing* process too. Figure 4.3 schematically illustrates the mechanism of difference-frequency generation in the usual fashion. Some part of the two input waves of frequencies ω_1 and ω_2 yield a difference-frequency ($\omega_3 = \omega_1 - \omega_2$) wave in the

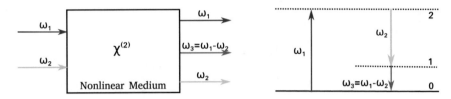

Fig. 4.3 Difference-frequency generation: **a** Schematic illustration **b** Energy band diagram

output. How it happens can be understood with the help of the band diagram shown in Fig. 4.3b. A photon of frequency ω_1 is absorbed and excites the atom to level 2. The excited atom in the energy level 2 is stimulated by the ω_2 field to decay to ground level via emission of two photons. The de-excitation from level 2 to 1 gives out frequency ω_2, and from level 1 to level 0 is accompanied by the release of a photon of frequency $\omega_3 = \omega_1 - \omega_2$, the difference frequency. It should be noted that the emission of ω_3 from level 1 to 0 may be spontaneous at first, but once the photons of ω_3 appear in the system, the transition $1 \rightarrow 0$ acquires stimulated nature [190, 191].

For the mathematical analysis of sum and difference generation, let us assume two input waves of frequencies ω_1 and ω_2, which add up to give the total optical field inside the nonlinear medium as

$$\tilde{E}(t) = E_1 e^{-i\omega_1 t} + E_2 e^{-i\omega_2 t} + c.c. \tag{4.23}$$

where $c.c.$ is $E_1^* e^{i\omega_1 t} + E_2^* e^{i\omega_2 t}$. The second-order polarization due to the total electric field $\tilde{E}(t)$ is

$$\tilde{P}^{(2)}(t) = \epsilon_0 \chi^{(2)} \tilde{E}^2(t) \tag{4.24}$$

Substituting Eq. 4.23 in Eq. 4.24, we get

$$\tilde{P}^{(2)}(t) = \epsilon_0 \chi^{(2)} [E_1 e^{-i\omega_1 t} + E_2 e^{-i\omega_2 t} + E_1^* e^{i\omega_1 t} + E_2^* e^{i\omega_2 t}]^2 \tag{4.25}$$

$$= \epsilon_0 \chi^{(2)} [(E_1 e^{-i\omega_1 t} + E_1^* e^{i\omega_1 t})^2 + (E_2 e^{-i\omega_2 t} + E_2^* e^{i\omega_2 t})^2 + \tag{4.26}$$

$$2(E_1 e^{-i\omega_1 t} + E_1^* e^{i\omega_1 t})(E_2 e^{-i\omega_2 t} + E_2^* e^{i\omega_2 t})] \tag{4.27}$$

$$= \epsilon_0 \chi^{(2)} [(E_1^2 e^{-i2\omega_1 t} + E_1^{*2} e^{i2\omega_1 t} + 2E_1 E_1^*) + (E_2^2 e^{-i2\omega_2 t} + \tag{4.28}$$

$$E_2^{*2} e^{i2\omega_2 t} + 2E_2 E_2^*) + 2(E_1 E_2 e^{-i(\omega_1+\omega_2)t} + E_1 E_2^* e^{-i(\omega_1-\omega_2)t} + \tag{4.29}$$

$$E_1^* E_2 e^{i(\omega_1+\omega_2)t} + E_1^* E_2 e^{i(\omega_1-\omega_2)t})] \tag{4.30}$$

$$= \epsilon_0 \chi^{(2)} [E_1^2 e^{-i2\omega_1 t} + E_2^2 e^{-i2\omega_2 t} + 2E_1 E_2 e^{-i(\omega_1+\omega_2)t} + \tag{4.31}$$

$$2E_1 E_2^* e^{-i(\omega_1-\omega_2)t} + c.c.] + 2\epsilon_0 \chi^{(2)} (E_1 E_2^* + E_1^* E_2) \tag{4.32}$$

The amplitude of second-order polarization for various frequency terms in Eq. 4.32 can be individually written and interpreted as

$$P(2\omega_1) = \epsilon_0 \chi^{(2)} E_1^2 = second\ harmonic\ of\ \omega_1 \qquad (4.33)$$

$$P(2\omega_2) = \epsilon_0 \chi^{(2)} E_2^2 = second\ harmonic\ of\ \omega_2 \qquad (4.34)$$

$$P(\omega_1 + \omega_2) = 2\epsilon_0 \chi^{(2)} E_1 E_2 = sum\ frequency\ term \qquad (4.35)$$

$$P(\omega_1 - \omega_2) = 2\epsilon_0 \chi^{(2)} E_1 E_2^* = difference\ frequency\ term \qquad (4.36)$$

$$P(0) = 2\epsilon_0 \chi^{(2)} (E_1 E_2^* + E_1^* E_2) = zero\ frequency\ term \qquad (4.37)$$

Though mathematically, all the above-obtained frequency components are probable to exist in the output, the one which actually exists depends on values of the input frequencies, which further depends on the energy band structure of the nonlinear medium employed.

4.3 Coupled Wave Equations

In the previous section, for the sum-frequency generation, we represented the time-varying electric fields for the two input waves as $\tilde{E}_1(t) = E_1 e^{-i\omega_1 t}$ and $\tilde{E}_2(t) = E_2 e^{-i\omega_2 t}$. On similar lines, the field of the sum-frequency $\omega_3 = \omega_1 + \omega_2$ can be written as $\tilde{E}_3(t) = E_3 e^{-i\omega_3 t}$. In general,

$$\tilde{E}_i(t) = E_i e^{-i\omega_i t}, \quad i = 1, 2, 3 \qquad (4.38)$$

where $i = 1, 2$ is meant for input frequencies and $i = 3$ is for sum-frequency component in the output. In this type of notation, it should be noted that the term $e^{-i\omega_i t}$ represents the time-dependent part and the remaining E_i is independent of time, but intrinsically contains the harmonic variation of the field w.r.t. space, which is a property of plane waves. The spatial term can be further expanded as

$$E_i = A_i e^{ik_i z} \qquad (4.39)$$

Hence, Eq. 4.38 becomes

$$\tilde{E}_i(t) = A_i e^{i(k_i z + \omega_i t)}, \quad i = 1, 2, 3 \qquad (4.40)$$

where $k_i = n_i \omega_i / c$ is the wave vector for the ith frequency and $n_i^2 = \epsilon^{(1)}(\omega_i)$ is the permittivity of the nonlinear medium for the ith frequency. Using this in Eq. 4.35, the amplitude of nonlinear polarization of the sum-frequency component is written as

$$P_3 = P(\omega_3) = 2\epsilon_0 \chi^{(2)} A_1 A_2 e^{i(k_1 + k_2)z} \qquad (4.41)$$

$$= 4\epsilon_0 d_{eff} A_1 A_2 e^{i(k_1 + k_2)z} \qquad (4.42)$$

where $d_{eff} = \chi^{(2)}/2$. The term d_{eff} hails from the tensor representation of nonlinear susceptibility. For a more detailed explanation of d_{eff}, the reader is advised to read Boyd [183].

Now, substituting Eqs. 1.44–1.46 in the wave equation for nonlinear medium, Eq. 1.19, we get a modified wave equation of the form

$$\frac{d^2 A_3}{dz^2} + 2ik_3 \frac{d A_3}{dz} = \frac{-4d_{eff}\omega_3^2}{c^2} A_1 A_2 e^{i(k_1+k_2-k_3)z} \tag{4.43}$$

Generally, the amplitude of the sum frequency varies minutely with respect to distance z, and hence the second derivative is very small compared to the first derivative and can be ignored. Thus, the wave equation is now reduced to

$$\frac{d A_3}{dz} = \frac{2i d_{eff}\omega_3^2}{k_3 c^2} A_1 A_2 e^{i\Delta kz} \tag{4.44}$$

where $\Delta k = k_1 + k_2 - k_3$ is the phase mismatch between the input and the output frequencies. This equation is referred to as a *coupled wave equation* [11, 183, 192–194], since it expresses a relation between the amplitudes of all the three frequencies. Here, the amplitude of ω_3 is being expressed in terms of those of ω_1 and ω_2. Similar equations can be written for the amplitudes A_1 and A_2, as shown below:

$$\frac{d A_1}{dz} = \frac{2i d_{eff}\omega_1^2}{k_1 c^2} A_3 A_2^* e^{-i\Delta kz} \tag{4.45}$$

$$\frac{d A_2}{dz} = \frac{2i d_{eff}\omega_2^2}{k_2 c^2} A_3 A_1^* e^{-i\Delta kz} \tag{4.46}$$

One needs to understand that conversion of ω_1 and ω_2 to ω_3 is not the only process taking place. In actuality, some of the power from ω_3 keeps converting back to ω_1 and ω_2, if there is a substantial phase mismatch. For the maximum conversion efficiency, the phase mismatch should be as low as possible, ideally $\Delta k = 0$. The role of phase matching (PM) in the conversion efficiency has been discussed in the next section.

4.4 Phase Matching

Phase matching is an important condition for three-wave mixing or second-harmonic generation processes. In a phase-matched nonlinear system, the atoms of the medium vibrate in phase with each other, and the emitted radiation is intense due to constructive interference [183, 195–198]. For a perfectly phase-matched system, $\Delta k = 0$, i.e., $k_1 + k_2 = k_3$.

Now, integrating Eq. 4.44 to obtain A_3 as a function of z, we get

$$A_3(z) = \frac{2i d_{eff} \omega_3^2 A_1 A_2}{k_3 c^2} \int e^{i\Delta kz} dz \tag{4.47}$$

$$= \frac{2i d_{eff} \omega_3^2 A_1 A_2}{k_3 c^2} \left(\frac{e^{i\Delta kz}}{i\Delta k} \right) + const. \tag{4.48}$$

where $const.$ is the constant of integration. Considering the boundary condition $A_3(0) = 0$ (no ω_3 at the input end), we get

$$const. = -\frac{2i d_{eff} \omega_3^2 A_1 A_2}{k_3 c^2} \left(\frac{1}{i\Delta k} \right) \tag{4.49}$$

Hence, eventually

$$A_3(z) = \frac{2i d_{eff} \omega_3^2 A_1 A_2}{k_3 c^2} \left(\frac{e^{i\Delta kL} - 1}{i\Delta k} \right) \tag{4.50}$$

Now from the field amplitudes, one can calculate the respective intensities as

$$I_i(z) = 2n_i \epsilon_0 c |A_i|^2, \quad i = 1, 2, 3 \tag{4.51}$$

Therefore, we get the intensity of the generated frequency as a function of distance and phase mismatch

$$I_3(z) = \frac{8 n_3 \epsilon_0 d_{eff}^2 \omega_3^4 |A_1|^2 |A_2|^2}{k_3^2 c^3} \left| \frac{e^{i\Delta kz} - 1}{\Delta k} \right|^2 \tag{4.52}$$

Substituting A_1 and A_2 in terms of I_1 and I_2 from Eq. 1.52, we get

$$I_3(z) = \frac{8 d_{eff}^2 \omega_3^2 I_1 I_2}{n_1 n_2 n_3 \epsilon_0 c^2} \left| \frac{e^{i\Delta kz} - 1}{\Delta k} \right|^2 \tag{4.53}$$

$$= \frac{8 d_{eff}^2 \omega_3^2 I_1 I_2}{n_1 n_2 n_3 \epsilon_0 c^2} z^2 \left[\frac{sin(\Delta kz/2)}{\Delta kz/2} \right]^2 \tag{4.54}$$

$$= \frac{8 d_{eff}^2 \omega_3^2 I_1 I_2}{n_1 n_2 n_3 \epsilon_0 c^2} z^2 sinc^2 \left(\frac{\Delta kz}{2} \right) \tag{4.55}$$

Figure 4.4 shows the variation of I_3 w.r.t. z in a nonlinear crystal for both zero and non-zero phase mismatch. For a non-zero phase mismatch (blue curve) Δk, $I_3 = 0$ at $z = 2n\pi/\Delta k$ where $n = 1, 2, 3 \ldots$. As marked in Fig. 4.4a by a dashed line, at $z = \pi/\Delta k$ the intensity becomes maximum and reduces beyond that. This distance is called the *coherence length* $L_{coh} = \pi/\Delta k$. The intensity of the generated wave shows sinusoidal distribution w.r.t. z. Hence to obtain maximum intensity of the sum frequency, the length of the crystal should be equal to the odd multiples of the coherence length.

Fig. 4.4 Significance of phase matching in nonlinear phenomena

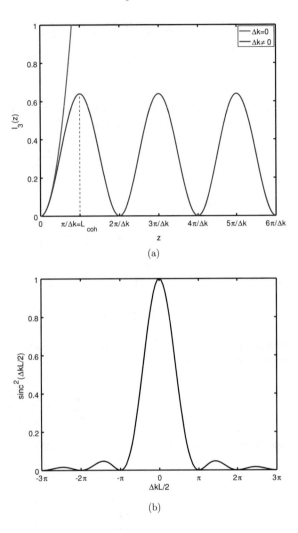

(a)

(b)

For a perfectly phase-matched situation ($\Delta k = 0$), the variation of intensity has been shown by the red curve. For $\Delta k = 0$, the sinc function becomes unity (Fig. 4.4b) and only the z^2 part survives, and hence the nature of the curve is parabolic. It can be observed that the intensity increases sharply with z in case of zero mismatch, since the coherence is always maintained. For crystal of a particular length L, the intensity I_3 only depends on Δ_k as

$$I_3(z) = \frac{8d_{eff}^2 \omega_3^2 I_1 I_2}{n_1 n_2 n_3 \epsilon_0 c^2} L^2 sinc^2 \left(\frac{\Delta k L}{2} \right) \tag{4.56}$$

Hence, the intensity now depends on the *sinc* part only and is maximum for $\Delta k = 0$, as illustrated in Fig. 4.4b. All the above discussion throws light at the one and only one fact, that is, Δk should be as low as possible because *the smaller the mismatch the greater the efficiency*. This sufficiently explains the significance of phase matching in nonlinear optics.

4.5 Achievement of Phase Matching

From the previous section, we know that the condition of phase matching is

$$\Delta k = k_1 + k_2 - k_3 = 0 \tag{4.57}$$

Since $k_i = n_i \omega_i / c$, where n_i is the refractive index of the medium for frequency ω_i, Eq. 4.57 can be written as

$$\frac{n_1 \omega_1}{c} + \frac{n_2 \omega_2}{c} = \frac{n_3 \omega_3}{c} \tag{4.58}$$

where $\omega_1 + \omega_2 = \omega_3$. In case of second-harmonic generation, $\omega_1 = \omega_2 = \omega$ and $\omega_3 = 2\omega$. Hence, for second-harmonic generation, Eq. 4.58 gives

$$n(\omega) = n(2\omega) \tag{4.59}$$

However, for sum-frequency generation we can rearrange Eq. 4.58 and get

$$n_3 = \frac{n_1 \omega_1 + n_2 \omega_2}{\omega_3} \tag{4.60}$$

$$n_3 - n_2 = \frac{n_1 \omega_1 + n_2 \omega_2}{\omega_3} - n_2 \tag{4.61}$$

$$= \frac{n_1 \omega_1 - n_2(\omega_3 - \omega_2)}{\omega_3} \tag{4.62}$$

$$= \frac{n_1 \omega_1 - n_2 \omega_1}{\omega_3} \tag{4.63}$$

$$= (n_1 - n_2)\frac{\omega_1}{\omega_3} \tag{4.64}$$

Equation 4.64 is the condition of *perfect phase matching* (PPM), which cannot be satisfied by normal dispersion where the refractive index exhibits increasing trend with increasing frequency, i.e., if $\omega_3 > \omega_2 > \omega_1$, then $n_3 > n_2 > n_1$. However, the condition can be met in the abnormal dispersion region, albeit it is a region of high absorption. Figure 4.5 shows the normal and anomalous dispersion regions in the refractive index versus frequency plot of silica. The data for the plot has been adapted from Palik [105]. The blue curve shows the real and the red curve shows

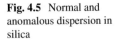
Fig. 4.5 Normal and anomalous dispersion in silica

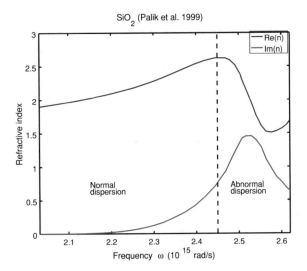

the imaginary part of the refractive index. The part of the graph where the refractive index increases with increasing frequency is the region of normal dispersion, where $n_3 > n_2 > n_1$ if $\omega_3 > \omega_2 > \omega_1$. And the region where this condition is not followed is called the anomalous dispersion region. The two regions have been separated in the graph by a vertical dashed line. As already stated, the normal dispersion region cannot satisfy Eq. 4.64 and cannot provide perfect phase matching (PPM). But, if at least one of the three refractive indices falls in the anomalous dispersion region, the said equation can be satisfied. However, the major challenge with working in the anomalous region is the high absorption on account of the high value of the imaginary part of the refractive index. Hence, the anomalous region may seem a solution at first, but on closer inspection, it turns out to be a bad idea. One needs a smarter method to achieve PM without being subject to significant losses. One such method is birefringence, in which the next section throws light on.

4.5.1 Birefringence

Certain crystals such as calcite, quartz, lithium niobate, KDP, etc. exhibit a strange behavior of splitting an incident ray of unpolarized light into two rays of different polarizations. This phenomenon is called *birefringence* or *double refraction* [10, 11, 13], and has been illustrated in Fig. 4.6. Figure 4.6a presents a ray diagram demonstrating the double refraction occurring inside a birefringent crystal and Fig. 4.6b illustrates the formation of two images of a single object as its consequence. The double refraction happens because birefringent crystals are asymmetric to different polarizations. In other words, different polarizations experience different refractive

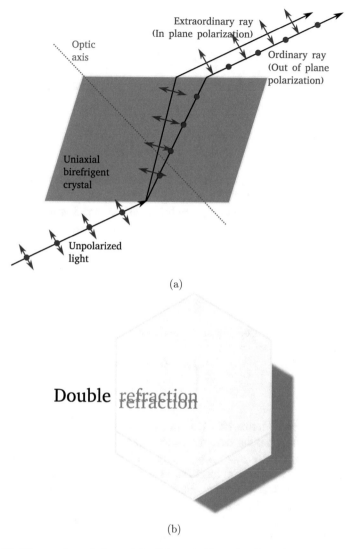

(a)

(b)

Fig. 4.6 Birefringence by a calcite crystal. **a** Schematic illustration, **b** two images of a single object as seen through a calcite crystal

indices of the medium. Such materials whose refractive index or permittivity is polarization dependent are called *anisotropic* materials.

We know that displacement vector $\mathbf{D} = \epsilon \mathbf{E}$, where ϵ is the permittivity and $\mathbf{E}$ is the electric field vector. In general, various components of the displacement vector can be written as

$$D_x = \epsilon_{xx} E_x + \epsilon_{xy} E_y + \epsilon_{xz} E_z \qquad (4.65)$$
$$D_y = \epsilon_{yx} E_x + \epsilon_{yy} E_y + \epsilon_{yz} E_z \qquad (4.66)$$
$$D_z = \epsilon_{zx} E_x + \epsilon_{zy} E_y + \epsilon_{zz} E_z \qquad (4.67)$$

In this way, permittivity is basically a tensor of rank two represented by

$$\epsilon = \begin{pmatrix} \epsilon_{xx} & \epsilon_{xy} & \epsilon_{xz} \\ \epsilon_{yx} & \epsilon_{yy} & \epsilon_{yz} \\ \epsilon_{zx} & \epsilon_{zy} & \epsilon_{zz} \end{pmatrix} \qquad (4.68)$$

where each component ϵ_{mn} of the matrix expresses the relation of mth component of $\mathbf{D}$ with nth component of $\mathbf{E}$. It can be shown that $\epsilon_{xy} = \epsilon_{yx}$, $\epsilon_{yz} = \epsilon_{zy}$ and $\epsilon_{zx} = \epsilon_{xz}$ [10, 13]. One can always choose a coordinate system such that all the off-diagonal terms are zero and only the diagonal terms are relevant. Then,

$$D_x = \epsilon_{xx} E_x \qquad (4.69)$$
$$D_y = \epsilon_{yy} E_y \qquad (4.70)$$
$$D_z = \epsilon_{zz} E_z \qquad (4.71)$$

In this way, the matrix of Eq. 4.68 is reduced to a diagonal matrix shown below:

$$\epsilon = \begin{pmatrix} \epsilon_{xx} & 0 & 0 \\ 0 & \epsilon_{yy} & 0 \\ 0 & 0 & \epsilon_{zz} \end{pmatrix} \qquad (4.72)$$

Now, if all the diagonal terms are equal, i.e., $\epsilon_{xx} = \epsilon_{yy} = \epsilon_{zz}$, the medium is said to be *isotropic*, and in any other case, it is *anisotropic*. Anisotropic media are further subdivided into *uniaxial* ($\epsilon_{xx} = \epsilon_{yy} \neq \epsilon_{zz}$) and *biaxial* ($\epsilon_{xx} \neq \epsilon_{yy} \neq \epsilon_{zz}$) media.

In Fig. 4.6, a uniaxial crystal has been assumed. When unpolarized light enters the medium it splits into two orthogonally polarized rays, labeled as *ordinary ray* and *extraordinary ray*. The ordinary ray (o-ray) travels with the same velocity in all directions while the extraordinary ray (e-ray) has different velocities in different directions. However, there is a particular direction, in which the e-ray travels at the velocity equal to that of the o-ray. This direction is called the *optic axis* of the crystal. Out of the two rays formed on account of birefringence, the one polarized normal to the plane containing the optic axis and the propagation vector becomes the o-ray, and the one whose polarization lies in the plane is the e-ray. The o-ray always sees the same refractive index n_o along all directions, while the e-ray experiences refractive index as n_o along the optic axis and n_e along the directions perpendicular to it. Along any other direction, the refractive index for e-ray lies between n_o and n_e. Depending on the value of n_e relative to n_o, a uniaxial crystal is classified as a *negative uniaxial* ($n_e < n_o$) crystal or a *positive uniaxial* crystal ($n_e > n_o$). In a positive uniaxial crystal, the velocity of the e-ray (v_e) is lesser than that of the o-ray v_o, along all directions except the optic axis, whereas in a negative crystal e-ray is faster than o-ray everywhere except along the optic axis, along which $v_e = v_o$. For better understanding and visualization, we have shown index ellipsoids of a negative and positive uniaxial crystal. An index ellipsoid is a 3D surface plot of refractive index along all the directions in Cartesian space. Figure 4.7a–b shows the index ellipsoids of a negative uniaxial crystal for both e-ray and o-ray, respectively. Here the optic axis has been assumed to be oriented along the z-axis, as per the convention. For e-ray, refractive index $n_e < n_o$ along x- and y-directions and $n_e = n_o$ along the z-direction. In the figure, the index ellipsoid of the negative crystal is blue in color and of prolate shape (like a rugby ball). The o-ray index ellipsoid is spherical indicating isotropic nature, since its refractive index is the same (n_o) in all directions. Figure 4.7c–d. presents the cross-sectional view of two index ellipsoids put together. It can be observed that index ellipsoid of the e-ray always remains inside that of the o-ray. Similarly, the index ellipsoids of a positive uniaxial crystal have been drawn and shown in Fig. 4.8. In this case, the index ellipsoid of e-ray is oblate in shape (like a pumpkin) and that of o-ray is again spherical. In a positive crystal, refractive index of e-ray is equal to n_o along the optic axis (the z-axis) and is equal to $n_e(>n_o)$ along x- and y-directions. Hence, in this case, it is the isotropic sphere of o-ray that always lies inside the ellipsoid of e-ray.

We now know refractive indices of both the types along the three axes, but cannot determine the refractive index of an extraordinary ray traveling in an arbitrary direction. In that endeavor, let us first think of an ellipse, as shown in Fig. 4.9. From our knowledge of geometry, we know that the equation of the ellipse is [199]

$$\frac{x^2}{a^2} + \frac{y^2}{b^2} = 1 \tag{4.73}$$

where a and b are the semi-major and semi-minor axes of the ellipse and (x, y) are the coordinates of an arbitrary point P on it. If the length of the line segment OP is r, then $x = r\cos\theta$ and $y = r\sin\theta$, where θ is the angle OP makes with respect to the x-axis. Then, Eq. 4.73 can be written as

$$\frac{r^2\cos^2\theta}{a^2} + \frac{r^2\sin^2\theta}{b^2} = 1 \tag{4.74}$$

or

$$\frac{1}{r^2} = \frac{\cos^2\theta}{a^2} + \frac{\sin^2\theta}{b^2} \tag{4.75}$$

Since r changes w.r.t. θ, we can write r as a function of θ as

$$\frac{1}{r^2(\theta)} = \frac{\cos^2\theta}{a^2} + \frac{\sin^2\theta}{b^2} \tag{4.76}$$

Let us analogously extend this equation to the case of index ellipsoids, following which the refractive index of the e-ray in an arbitrary direction is given by

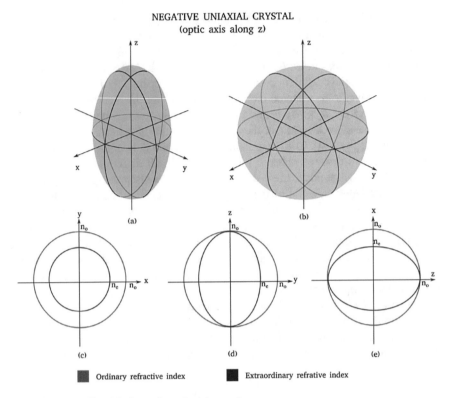

NEGATIVE UNIAXIAL CRYSTAL
(optic axis along z)

Fig. 4.7 Index ellipsoid of negative uniaxial crystals

$$\frac{1}{n^2(\theta)} = \frac{cos^2\theta}{n_0^2} + \frac{sin^2\theta}{n_e^2} \tag{4.77}$$

Here, θ is the angle between the direction of propagation (i.e., the velocity vector) and the optic axis, and hence a has been replaced by n_o and b has been replaced by n_e. For $\theta = 0$, $n = n_o$ and for $\theta = \pi/2$, $n = n_e$, which absolutely satisfies the axial propagation. Hitherto, we have rigorously discussed the concept of birefringence, which is sufficient to understand its significance in facilitating the reduction of phase mismatch as discussed below.

4.5.1.1 Phase Matching by Means of Birefringence

Birefringence can facilitate phase matching in a very subtle way. The idea is to have the fundamental frequency and the second harmonic to be orthogonally polarized, such that the fundamental frequency (ω) travels as the ordinary wave and the second harmonic travels as the extraordinary wave [200, 201]. The refractive index for the

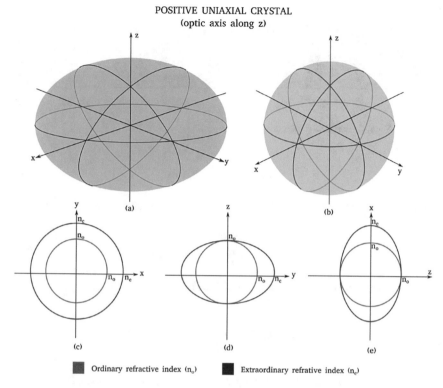

POSITIVE UNIAXIAL CRYSTAL
(optic axis along z)

Fig. 4.8 Index ellipsoid of positive uniaxial crystals

Fig. 4.9 An ellipse

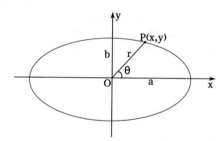

fundamental frequency will be n_o and for the second-harmonic frequency will be $n(\theta)$ depending on the angle (θ) between the velocity vector and the optics axis. We know that for $\Delta k = 0$ in case of second-harmonic generation

$$n(\omega) = n(2\omega) \tag{4.78}$$

which implies that

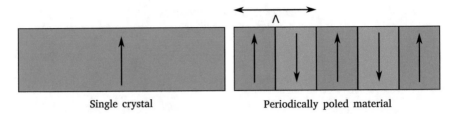

Single crystal   Periodically poled material

Fig. 4.10 Second-harmonic generation

$$n_o(\omega) = n_e(2\omega, \theta) \tag{4.79}$$

Using this condition, Eq. 4.77 can be rewritten for SH frequency as

$$\frac{1}{n_o^2(\omega)} = \frac{cos^2\theta}{n_0^2(2\omega)} + \frac{sin^2\theta}{n_e^2(2\omega)} \tag{4.80}$$

Solving this equation, we get the value of θ which ensures $\Delta k = 0$

$$\theta = sin^{-1}\left[\frac{\frac{1}{n_o^2(\omega)} - \frac{1}{n_o^2(2\omega)}}{\frac{num}{den} - \frac{n_e^2(\omega)}{n_o^2(\omega)}}\right] \tag{4.81}$$

Hence, we eventually get that if light propagates at the angle θ (given by the above equation) w.r.t. the optic axis, there will be no phase mismatch between the fundamental and the second-harmonic wave. In this way, birefringence comes to our refuge and serves the purpose of phase matching. By this, we have gathered sufficient knowledge about nonlinear optics needed to understand the significance of zero-index metamaterial in this area.

4.5.2 Quasi-phase Matching

The crystals with low or absolutely no birefringence are incapable of compensating dispersion. In such a situation, an artificial medium is specially fabricated from a single crystal such that the orientation of the c-axis inside the new medium is inverted alternatively, as shown in the figure below, along the length of the material in the intended direction of wave propagation. Such type of material is called a *periodically poled material* (see Fig. 4.10). The periodicity $\Lambda = 2L_{coh}$, where L_{coh} is the coherence length. This technique of dispersion compensation using a periodically poled material is called *quasi-phase matching* [202–205]. Periodic inversion of the c-axis results in periodic toggling of the sign of d_{eff}, which results in dispersion compensation.

4.6 Second-Harmonic Generation

From the previous sections, we understand that a crystal with sufficiently high nonlinear susceptibilities can convert a part of the incident fundamental frequency into the second-harmonic frequency and that the efficiency of conversion depends, to a great extent, on the length of the crystal. Armstrong et al. [183, 206] presented a beautiful mathematical analysis of the exchange of power between the two frequency as they propagate through the crystal. They derived exact solutions to the coupled wave equations for both the perfect (PPM) and the imperfect phase matching (IPM) cases and represented the two field amplitudes as a function of distance. The mathematical analysis by Armstrong et al. has been discussed below.

Firstly, let us consider a nonlinear crystal of length L, as shown in Fig. 4.11. Light is assumed to be propagating along the z-direction. As light propagates through the crystal and fundamental frequency converts into the second-harmonic frequency, the amplitudes of the two waves vary as [183]

$$A_1 = \left(\frac{I}{2n_1 \epsilon_0 c} \right)^{1/2} u_1 e^{i\phi_1} \tag{4.82}$$

$$A_2 = \left(\frac{I}{2n_2 \epsilon_0 c} \right)^{1/2} u_2 e^{i\phi_2} \tag{4.83}$$

where u_1 and u_2 are the normalized field amplitudes of the fundamental and the second-harmonic waves, respectively, $I = I_1 + I_2$ is the total intensity of the two waves, and the rest of the symbols have their usual meaning. As a consequence of the law of conservation of energy, I remains constant. Hence,

$$u_1^2 + u_2^2 = 1 \tag{4.84}$$

In their mathematical analysis, instead of distance z, Armstrong et al. use a normalized distance parameter ζ given as

$$\zeta = z/l \tag{4.85}$$

where

$$l = \left(\frac{2n_1^2 n_2}{\epsilon_0 c I} \right)^{1/2} \frac{c}{2\omega_1 d_{eff}} \tag{4.86}$$

is the characteristic distance for the exchange of power between the two frequencies. They also introduced a relative phase parameter θ given by

$$\theta = 2\phi_1 - \phi_2 + \Delta k z \tag{4.87}$$

which encompasses the phase terms of the two field amplitudes as well as a normalized phase mismatch parameter

$$\Delta s = \Delta k l \tag{4.88}$$

Writing the coupled wave equation in terms of u_1 and u_2, we get the following equations:

$$\frac{du_1}{d\zeta} = u_1 u_2 sin\theta \tag{4.89}$$

$$\frac{du_2}{d\zeta} = -u_1^2 sin\theta \tag{4.90}$$

$$\frac{\theta}{d\zeta} = \Delta s + \frac{cos\theta}{sin\theta} \frac{d}{d\zeta} (\ln u_1^2 u_2) \tag{4.91}$$

Solving the above equations for a perfectly, phase-matched ($\Delta s = 0$) situation yields simple solutions as follows:

$$u_1(\zeta) = sech\zeta \tag{4.92}$$

and

$$u_2(\zeta) = tanh\zeta \tag{4.93}$$

Figure 4.12 graphically illustrates the growth of the second-harmonic field and the reduction of the fundamental field with propagation distance. The development of the second-harmonic occurs at the cost of the fundamental wave, i.e., power flows from the fundamental wave to second-harmonic wave. For the perfect phase match, the power transfer or, more accurately, the conversion efficiency is 100% as shown. Better visualization of the growing u_2 and the diminishing u_1 has been presented in Fig. 4.13, as an aid to the imagination. In general, efficiency of second-harmonic generation at the end of the crystal of length L is given as

$$\eta = \frac{u_2^2(L)}{u_1^2(0)} \tag{4.94}$$

To understand the field variation inside the crystal with imperfect phase matching, Armstrong et al. provided an approximate formula according to which the second-harmonic amplitude for different values of mismatch is given by

$$p_2 = \frac{4p_1(0)}{\Delta s} sin\left(\frac{\Delta s\zeta}{2}\right) \tag{4.95}$$

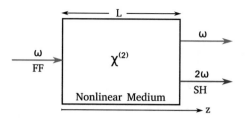

Fig. 4.11 Second-harmonic generation

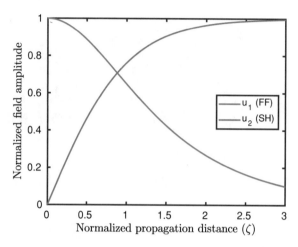

Fig. 4.12 Second-harmonic generation with perfect phase matching, according to Armstrong et al.

The variation of the field, according to Eq. 4.95, is shown in Fig. 4.14, in which it can be observed that the second-harmonic field does not increase continuously till a saturation value, as it happened in the case of perfect phase matching. Instead, the field amplitude varies periodically. The greater the value of Δs the lesser the second-harmonic amplitude and the smaller the period. It is necessary to mention here that the fundamental amplitude remains practically constant throughout the crystal, as a consequence of very low conversion efficiency. Hence, it is desired that mismatch should be as low as possible so that the phase velocities of the fundamental and the second-harmonic waves are approximately equal, and efficient flow of power from the former to the latter can take place.

As discussed in earlier sections, PPM is not possible in case of normal dispersion where $n(2\omega) > n(\omega)$ since phase velocities of the two waves are unequal. The most well-acknowledged methods to compensate dispersion are by using birefringent crystal and quasi-phase matching (QPM), as discussed before. In addition to these, a very innovative route to achieve a phase-matched condition has surfaced during recent years, which has the potential to revolutionize the nonlinear optics, i.e., *via zero refractive index metamaterials*. The idea behind their low mismatch allowance

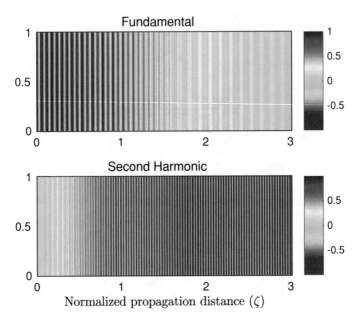

Fig. 4.13 Second-harmonic generation with perfect phase matching, according to Armstrong et al.

Fig. 4.14 Second-harmonic
generation with imperfect
phase matching, according to
Armstrong et al.

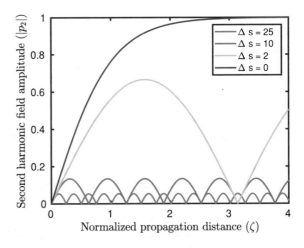

is that the individual propagation constants $k_1(=n_1k_0)$ and $k_2(=n_2k_0)$ get diminished
on account of close to zero values of n_1 and n_2, and consequently Δk gets diminished too. The effectiveness of zero-index media in boosting the second-harmonic
generation has been discussed in the next section in detail.

4.6.1 SHG in Zero-Index Medium

From the previous sections, we know that the momentum mismatch or phase mismatch Δk in the second-harmonic generation is given by

$$\Delta k = 2k_1 - k_2 \tag{4.96}$$

$$= 2n_1\frac{\omega}{c} - n_2\frac{2\omega}{c} \tag{4.97}$$

$$= 2(n_1 - n_2)\frac{\omega}{c} \tag{4.98}$$

Now, there are two ways of reducing Δk. One is to have $n_1 \approx n_2$, which is prevented by normal dispersion. The other is by reducing both n_1 and n_2, close to zero, which can be achieved by employing zero-index metamaterials.

To showcase the efficiency of zero-index metamaterials in enhancing the second-harmonic field, we present below a comparative study between a homogeneous silicon medium and a silicon-based zero-index metamaterial. Let us first analyze the case of a zero-index medium. We have chosen a rods-in-air-type metamaterial whose zero-index behavior is a manifestation of accidental degeneracy-induced Dirac cone. It is similar to the one proposed by Huang et al. [27], whose geometry has already been shown in Fig. 2.7 Sect. 2.5 Chap. 2. The rods are made up of silicon and the radius r of each rod is 0.2a, where a is the periodicity of the square lattice. The band structure has been computed by *plane wave expansion method* (PWEM) [42], in which a constant value of relative permittivity ($\epsilon_r = 12.5$) has been chosen for the silicon domains and the surrounding medium is air. The band diagram with ten bands has been shown in Fig. 4.15. The zero-index character of this metamaterial around the Dirac cone has already been explained in Chap. 2. *Now, how to study second-harmonic generation in this zero-index metamaterial?*. Firstly, the Dirac frequency $\omega a/2\pi c = 0.542 = a/\lambda$, marked by the red dot, has been chosen as the fundamental frequency, whose corresponding second-harmonic frequency has been located in the band diagram at $\omega a/2\pi c = 1.089 = a/\lambda$, indicated by the blue dot. It can be noticed that both these frequencies correspond to zero wave vector, being at the center of the Brillouin zone (Γ point). Depending upon the desired fundamental frequency (or wavelength) periodicity a is decided. We choose periodicity $a = 840$ nm, so that the two frequencies translate into wavelengths $\lambda_1 = 1550$ nm (fundamental) and $\lambda_2 = 771$ nm $\approx \lambda_1/2$ (second harmonic). Using the band structure with some simple mathematics, the phase velocities $v_p = \omega/k$ have been calculated, followed by the calculation of refractive index as $n = c/v_p$, at different frequencies (or equivalently at different free-space wavelengths). The refractive index obtained as the function of wavelength, around both the fundamental and the second-harmonic frequencies, has been shown in Fig. 4.16. According to these two graphs, the refractive index at λ_1 is $n_1 = -0.0036$, and at λ_2 is $n_2 = 0.0040$. Using these in Eq. 4.98, we get $\Delta k = -6.162 \times 10^4$ m^{-1}. Now, to know if this value is substantially and sufficiently low, it needs to be compared with a homogeneous silicon medium. The

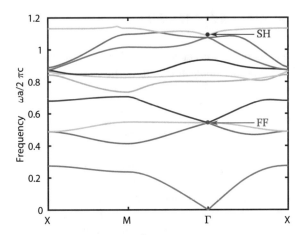

Fig. 4.15 Second-harmonic generation with imperfect phase matching, according to Armstrong et al.

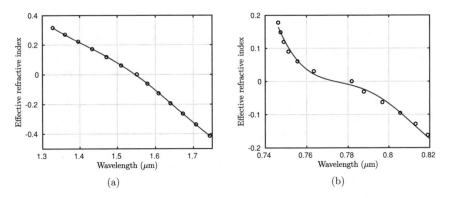

Fig. 4.16 Refractive index of the metamaterial determined around **a** the fundamental and **b** the second-harmonic frequencies. The markers indicate the actual values obtained from the graph and the blue curved are the fourth-order polynomial fit

refractive index of silicon at λ_1 is $n_1 = 3.477$ and at λ_2 is $n_2 = 3.714$ (Palik 1999). In the case of Si, we get $\Delta k = -1.921 \times 10^6 \text{ m}^{-1}$, which is more than 30 times than that in the case of the zero-index metamaterial. Hence, we see that the zero-index metamaterial allows drastically low phase mismatch compared to the homogeneous medium of the same material.

4.6.1.1 Time-Domain Computation

Lowering the phase mismatch results in the enhancement of the second-harmonic field. But before proceeding to the calculation of the field, one first needs to ensure

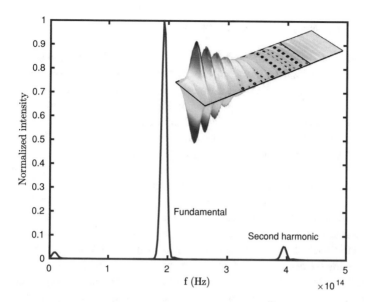

Fig. 4.17 Second-harmonic generation with imperfect phase matching, according to Armstrong et al.

whether the second generation can take place inside the zero-index metamaterial under consideration. For this purpose, the ZIM has been evaluated in the time domain using COMSOL Multiphysics. The ZIM slab used was ten periods long and five periods wide. A 1 W pulse of delay $t_0 = 120$ fs and width $\Delta t = 50$ fs was allowed to pass through the ZIM completely. The minimum spot size of the beam was

$$w0 = \sqrt{\frac{L\lambda_1}{2\pi n_0}} \tag{4.99}$$

where $L = 10a$ is the length of the ZIM, $\lambda_1 = 1.55\,\mu$m is the fundamental wavelength, and $n_0 = 1$ is the refractive index of air. The nonlinear susceptibility of silicon has been taken to be 15 pm/V. The Fourier transform of the field obtained at the output probe results in the following spectrum. In Fig. 4.17, the prominent fundamental peak is observed at $f_1 = 192.5$ THz ($\equiv \lambda_1 = 1550$ nm), and a tiny second-harmonic peak is present at 394.4 THz ($\approx 2f_1$). The tiny extra peak near zero is of the difference frequency and is irrelevant. This analysis shows the potential of the considered zero-index metamaterial for the second-harmonic generation. Now the next step should be the calculation of the individual fields and to determine the conversion efficiency.

4.6.1.2   Frequency-Domain Computation

To calculate the fundamental and the second-harmonic fields inside the zero-index medium and the homogeneous silicon medium, frequency-domain computations have been performed in COMSOL multiphysics. The designs of the computation regions in the two cases have been shown in Fig. 4.18. The ZIM region is ten periods long and five periods wide (same as earlier) and has air space of 2.5 μm on both the sides along the length. An intense plane wave of power 300 MW/m² is launched toward the ZIM from the air region, as indicated in the figure. The reason for such high power is that the nonlinear susceptibilities can come into play only in intense fields. The nonlinear second-order susceptibility of silicon used here is $\chi^{(2)} = 15$ pm/V [207]. In the case of homogeneous medium, an equivalent Si slab of the same size $10a \times 5a$ has been taken with the same type of air regions on both sides and the same input power. The results of the computation have been shown in Figs. 4.19 and 4.20. Figure 4.19a–c presents the case of rods. Figure 4.19a and b presents the 2D plots showing the distribution of z-component of electric field E_z, inside and outside the metamaterial, for both the fundamental and the second-harmonic waves. The fundamental wave appears to fade and the second-harmonic appears to grow, which satisfies the expectation. For a closer inspection of the electric field inside the ZIM, 1D plots have been drawn and shown in Fig. 4.19c. The red and blue curves represent the fundamental and second-harmonic waves, respectively. Since the refractive index of the ZIM is close to zero, the wavelength inside it is very large. Hence, as a trough enters the ZIM, it gets stretched along the length of the crystal (dashed red). The tiny waviness that one sees inside the ZIM arises due to the periodic nature of the structure and must not be used to predict the fundamental wavelength inside the ZIM. Similarly, in the case of the second harmonic, a crest gets stretched inside the ZIM, as shown in Fig. 4.19c by the dashed-blue curve. The overriding wave appears due to the periodic nature of the structure. It can be noticed that number of overriding crests or troughs inside the ZIM is ten in each of the two graphs, as the ZIM is ten periods long. The maximum amplitude of the fundamental field inside the ZIM is of the order -5×10^5 V/m while that of second harmonic is around 2 V/m and the efficiency of SHG is determined as

$$\eta = \left(\frac{Amplitude\ of\ the\ fundamental\ wave\ before\ entering\ the\ ZIM}{Amplitude\ of\ the\ SH\ wave\ after\ exiting\ the\ ZIM} \right)^2 \tag{4.100}$$

$$= \left(\frac{2.37}{4.75 \times 10^5} \right)^2 \tag{4.101}$$

$$= 2.5 \times 10^{-11} \tag{4.102}$$

It summarizes everything we obtained by numerically analyzing the ZIM. Though the efficiency is poor, to know if it is better or worse, let us next scrutinize the results for homogeneous Si medium displayed in Fig. 4.20. The 2D plots of Fig. 4.20a and

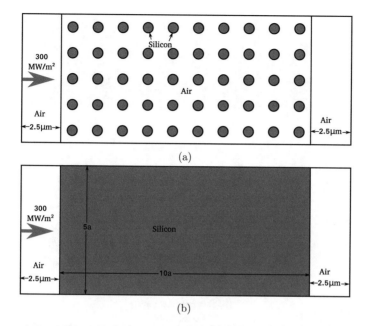

Fig. 4.18 Computational cells for **a** ZIM and **b** silicon slab

b show the distribution of E_z of fundamental and the second harmonic in the computational cell. The amplitude of the fundamental wave is the same throughout the slab, but that of the second harmonic shows periodic variation. The same feature is observed in the 1D plots of the two fields in Fig. 4.20c. The periodic variation is attributed to the non-zero mismatch, owing to the normal dispersion of the pure silicon. The maximum of the periodically varying amplitude of the SH field is 0.3 V/m, while the amplitude of FF remains the same 2×10^5 V/m. The efficiency in this case is

$$\eta = \left(\frac{\text{Amplitude of the fundamental before entering the Si slab}}{\text{Amplitude of the SH after exiting the Si slab}} \right)^2 \qquad (4.103)$$

$$= \left(\frac{0.265}{4.75 \times 10^5} \right)^2 \qquad (4.104)$$

$$= 3.11 \times 10^{-13} \qquad (4.105)$$

Since both the devices were fed with the same input power, the amplitude of the excitation input field is the same for both of them, i.e., 4.75×10^5 V/m. Comparing the two efficiencies, we observed that with ZIM the efficiency is more than 80 times of that with pure Si slab. And it's noteworthy that this improved efficiency has been achieved with just 12.5% amount of the nonlinear material, since the silicon filling fraction in ZIM is $\pi(0.2a)^2/a^2 = 0.1257$. It is solely the role of zero refractive

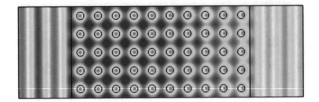

(a) Fundamental

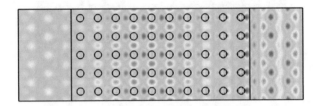

(b) Second harmonic

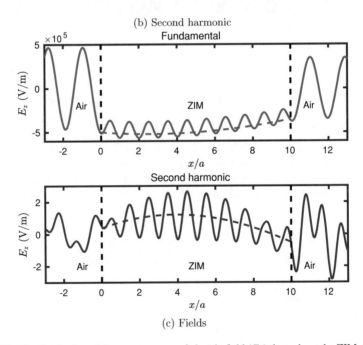

(c) Fields

Fig. 4.19 The distribution of the z-component of electric field (E_z) throughout the ZIM. One can notice **a** a fading fundamental field and **b** a growing second-harmonic field. **c** The variation of the two fields with distance

(a) Fundamental

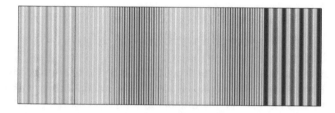

(b) Second harmonic

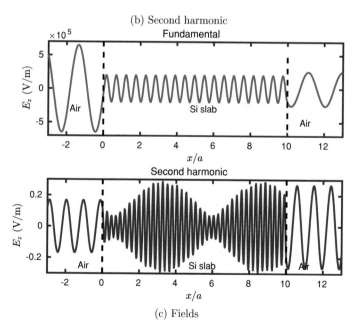

(c) Fields

Fig. 4.20   The distribution (E_z) throughout the Si slab. The fundamental field (**a**) and the second-harmonic field (**b**). The variation of the two fields with distance (**c**)

index in improving the efficiency drastically. We believe that the above discussion presents the significance of zero-index metamaterial in the SHG in a sufficiently rigorous manner. Next, we tackle another important nonlinear phenomenon called *self-focusing* and check if it can be enhanced with ZIM or not.

4.7 Intensity-Dependent Refractive Index

From the initial sections of this chapter, we know that total polarization

$$\hat{P}(t) = \epsilon_0(\chi^{(1)}\tilde{E}(t) + \chi^{(2)}\tilde{E}^2(t) + \chi^{(3)}\tilde{E}^3(t) + \cdots) \tag{4.106}$$

$$= \epsilon_0\tilde{E}(t)(\chi^{(1)} + \chi^{(2)}\tilde{E}(t) + \chi^{(3)}\tilde{E}^2(t) + \cdots) \tag{4.107}$$

Since we are talking about intensity dependence, here the third-order susceptibility becomes the most important because it has $|E|^2$ in multiplication with it, and we already know that intensity is proportional to the square of the electric field. For the sake of simplicity, let us drop $\chi^{(2)}$ from the above equation and consider the contribution of the third-order nonlinear susceptibility $\chi^{(3)}$ only [208–210]. Then, we can write

$$P^{TOT} = P + P^{NL} \tag{4.108}$$

$$= \epsilon_0 E(\chi^{(1)} + \chi^{(3)}|E|^2 + \cdots) \tag{4.109}$$

where $|E|^2$ represents the intensity factor and E is the complex amplitude of the electric field. Replacing $\chi^{(1)}$ by $\epsilon^{(1)} - 1$, we get

$$P^{TOT} = P + P^{NL} \tag{4.110}$$

$$= \epsilon_0 E(\chi^{(1)} + \chi^{(3)}|E|^2 + \cdots) \tag{4.111}$$

Therefore, the relative permittivity is

$$\epsilon = 1 + (\epsilon^{(1)} - 1 + 3\chi^{(3)}|E|^2) \tag{4.112}$$

$$= \epsilon^{(1)} + 3\chi^{(3)}|E|^2 \tag{4.113}$$

or

$$n^2 = \epsilon^{(1)} + 3\chi^{(3)}|E|^2 \tag{4.114}$$

hence

$$n = \sqrt{n_0^2 + 3\chi^{(3)}|E|^2} \tag{4.115}$$

where $n_0 = \sqrt{\epsilon^{(1)}}$ is the linear refractive index. In order to introduce nonlinear refractive index, we write the above equation as

$$n = \sqrt{n_0^2 + 2n_0 n_2 I} \tag{4.116}$$

where $I = 2Re\{n_0\}\epsilon_0 c|E|^2$ is the intensity of the optical field and

$$n_2 = \frac{3\chi^{(3)}}{4n_0 Re\{n_0\}\epsilon_0 c} \tag{4.117}$$

is the nonlinear refractive index as defined by Boyd and Reshef et al. [121, 211]. Equation 4.114 is the most general and accurate method of calculating the refractive index, including both the linear and the nonlinear contributions. If the contribution of the nonlinear part is relatively small, i.e., $\frac{2n_2 I}{n_0} << 1$,

$$n = n_0\sqrt{1 + \frac{2n_2 I}{n_0}} \approx n_0\left[1 + \frac{1}{2}\left(\frac{2n_2 I}{n_0}\right) + \cdots\right] \tag{4.118}$$

In most natural materials, $\frac{2n_2 I}{n_0}$ is very small, and hence all the higher terms can be neglected and total refraction can be written as

$$n = n_0 + n_2 I \tag{4.119}$$

Typically, silicon has $n_0 = 3.44$, $\chi(3) = 2.45 \times 10^{-19}$ m^2/V^2 and $n_2 = 5.52 \times 10^{-18}$ m^2/W at $\lambda = 1.55\,\mu$m [212].

4.7.1 Self-focusing

The intensity dependence of the refractive index results in an interesting phenomenon called *self-focusing*, which occurs for a Gaussian beam [213–216]. A Gaussian beam has the maximum intensity at the center, which reduces gradually and symmetrically away from the center. As a result, a medium with positive n_2, on being exposed to a Gaussian beam, acquires Gaussian refractive index profile, i.e., maximum at the center and gradually reducing in the radial direction. Normally, a low-intensity beam diverges on account of diffraction, but for a high-intensity beam, the induced Gaussian refractive index profile behaves as a converging lens, canceling the effect of diffraction. Here comes a significant parameter called *critical power*, given by [183]

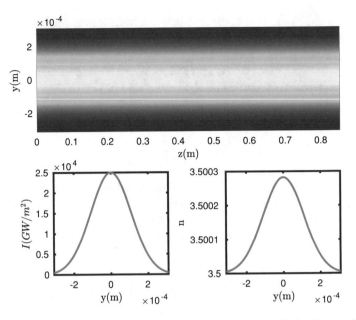

Fig. 4.21 The visualization of a Gaussian beam (top), its intensity distribution (bottom-left), and the refractive index profile created due to nonlinear effects (bottom-right)

$$P_{cr} = \frac{\pi(0.61)^2\lambda_0^2}{8n_0n_2} \tag{4.120}$$

$$= \frac{\pi(0.61)^2\lambda_0^2n_0\epsilon_0c}{12\chi^{(3)}} \tag{4.121}$$

Now, if the beam power P is equal to P_{cr}, the convergence due to nonlinearity exactly cancels the divergence due to diffraction. In this situation, the beam remains perfectly parallel throughout the medium with constant spot size, and the phenomenon is called *self-trapping* [217–219]. Whereas if the beam power is more than P_{cr}, the nonlinear effect overpowers the diffraction effect, and the beam ends up being converged to a small spot. This phenomenon is called *self-focusing*, and the distance of the focus point from the entrance boundary is called self-focusing distance, given by

$$z_{sf} = \frac{2n_0w_0^2}{\lambda_0}\frac{1}{\sqrt{P/P_{cr}-1}} \tag{4.122}$$

and the corresponding *self-focusing angle* is given by

$$\theta_{sf} = \sqrt{2n_2I/n_0} \tag{4.123}$$

By means of self-focusing, extremely intense laser spots can be obtained. It is important to mention here that besides the above two, there can be a third case as well. If the beam power is very very large compared to the critical power ($P \gg P_{cr}$), the beam breaks up into several small filaments, each one of which contains the same power approximately equal to P_{cr}. Out of these three, the self-focusing shall remain the topic of concern and discussion in this section. So far, we have acquired the basic understanding of self-focusing. Next, we shall discuss how it is pertinent to zero-index metamaterials.

4.7.2 Intensity Dependence of a Zero Linear Refractive Index Medium

Equation 4.119 is the most commonly used method to determine the total refractive index of a nonlinear medium because one mostly deals with natural materials and/or their alloys. However, the scenario is completely changed, challenging the common understanding, when the linear refractive index (n_0) vanishes, as in a zero-index metamaterial. In a ZIM, if $n_0 \to 0$, then $n_2 \to \infty$, $P_{cr} \to 0$, $z_{sf} \to 0$. For a non-linear refractive index tending to infinity, the expansion shown in Eq. 4.118 and the consequent Eq. 4.119 are rendered invalid [121]. As explained by Reshef et al. [121], $n_2 I$ does not remain a small perturbation in the refractive index and rather dominates the linear term n_0. Hence, in the case of zero refractive index medium, it is best to use Eq. 4.114 for accurate calculation of n, since it involves $\chi^{(3)}$ and E which are independent of n_0. The anomaly of Eq. 4.119 is illustrated in Fig. 4.23 citing the case of indium tin oxide (ITO) thin film laid on a glass substrate, illuminated at its zero epsilon wavelength 1240 nm [100, 121]. Here $\chi^{(3)}$ of ITO has been taken to be 2.16×10^{-18} m^2/V^2 [220], and $n_0 = 0.44$ $n_2 = 0.016$ cm^2/GW [121]. There is a noticeable difference between the curves corresponding to Eqs. 4.119 and 4.114 at higher intensities, which yields inaccurate values of n for wavelength corresponding to near-zero refractive index. Hence, under these circumstances, it is advisable to use Eq. 4.114. ITO is an attractive candidate for nonlinear optical applications on account of its considerably high nonlinear susceptibility as well as its CMOS compatibility

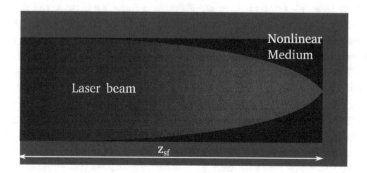

Fig. 4.22 Schematic illustration of self-focusing

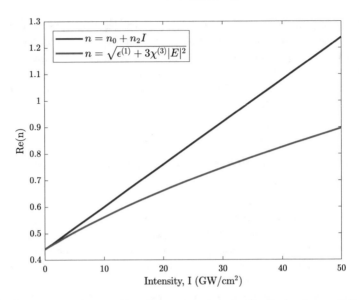

Fig. 4.23 Visualization of a Gaussian beam (top), its intensity distribution (bottom-left), and the refractive index profile created due to nonlinear effects (bottom-right)

proves it suitable for integrated optics. The potential of ITO has been discussed in detail in the next section.

4.7.2.1 Self-focusing in a Zero-Index Medium

The expressions of all the relevant parameters described above include low-intensity refractive index n_0, which either appears in the numerator or the denominator. For example, in the case of silicon, the values of P_{cr}, z_{sf}, and θ_{sf} come out to be 8874 W, 0.1085 m, and 0.012707 rad, respectively, at wavelength $\lambda = 1.55\,\mu$m and for beam width $w_0 = 100\,\lambda$ and $n_0 - 3.5$. It implies that in a zero-index medium, the nonlinear refractive index becomes so large that self-focusing is achieved at very low power and very short distance. We numerically analyzed the self-focusing in a silicon medium and a zero-index medium, whose results have been shown in Fig. 4.24. Firstly, Fig. 4.24a shows the schematics of self-focusing with the definition of z_{sf}. Figure 4.24b and c, respectively, shows the distribution of intensity and electric field E_z for self-focusing in silicon, whereas Fig. 4.24d and e shows these quantities fo zero-index metamaterial. In the electric field plot of silicon, the narrowing of the beam and focusing of the wavefronts are visible, but in that of the ZIM, the wavefronts are difficult to recognize because inside the ZIM the wavelength has become enormous, albeit the focus is recognizable in the intensity plots for both the cases. We determined that while for Si medium $z_{sf} = 0.108$ m, in ZIM it is reduced drastically to 74.1 μm, i.e., by the factor 1/1457. The employment of zero-index metamaterial

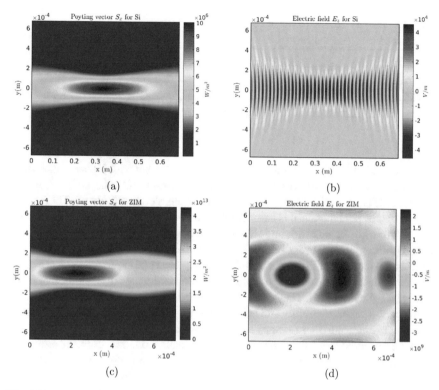

Fig. 4.24 Numerical demonstration of self-focusing for a silicon slab (**a**)–(**b**) and silicon-rods-based zero-index metamaterial (**c**)–(**d**)

reduced the focusing distance from centimeter scale to micrometer scale making the focusing achievable in integrated optics-based devices. Similarly, critical power has also reduced from 8875 W in the case of Si to 6 W for ZIM, by the factor 1/1480, which again makes it more suitable for integrated optics. In this way, by means of zero-index media intense laser spots can be obtained in low-power microscopic devices.

4.8 Nonlinear Response of Indium Tin Oxide Thin Film Near Bulk Plasma Frequency

Indium tin oxide (ITO) is a transparent conducting oxide (TCO) synthesized in the form of thin films deposited on a glass substrate by RF sputtering method. The target used is usually consisting of 90% indium oxide (In_2O_3) and 10% tin oxide (SnO_2) by weight. ITO has attracted tremendous interest because of its conductivity and optical transparency, and being CMOS-compatible, it finds applications in numerous

Table 4.1 The nonlinear properties of ITO thin films

Film properties (size: 1×1 cm)	Sample 1	Sample 2	Sample 3	Sample 4
Thickness (nm)	118	130	138	133
Carrier conc. $N_d (\times 10^{20}$ cm$^{-3})$	4.0	4.4	5.4	5.8
$n_2 \times 10^{-5}$ cm^2/GW at $\lambda = 720$ nm	1.6	2.1	2.6	4.1
$Re\{\chi^{(3)}\} (\times 10^{-13}$ esu) at $\lambda = 720$ nm	15	21	25	40

optoelectronic devices. ITO is an *n-type* semiconductor but with high conductivity it is close to metals. For example, while the conductivity is of the order of $10 \times^7$ S/m for silver, for ITO, it is of the order of $10 \times^5$ S/m. Though conductivity of ITO is not as high as that of pure metals, it is comparable to certain metal alloys, such as nichrome ($9 \times^5$ S/m), while being astronomical as compared to that of common semiconductors such as silicon, whose order of conductivity is $10 \times^{-3}$ S/m. High conductivity and optical transparency of ITO have been largely explored and exploited [221–224], but in 2006 Elim et al. performed an interesting and unprecedented investigation of ITO's nonlinear behavior [225]. They excellently studied the effect of carrier concentration on the nonlinear susceptibility and refractive index of various ITO films deposited on a glass substrate, and their findings have been presented in the table. From Table 4.1, it can be observed that by increasing carrier concentration, the nonlinear properties of ITO can be enhanced. The observed values of the nonlinear refractive index for ITO are drastically higher than almost all other materials, whether they are crystals, glasses, polymers, liquids, or even nanoparticles (see Table 4.1.2 in Boyd [183]). This high nonlinearity and CMOS compatibility make ITO a very lucrative candidate for on-chip nanophotonic applications.

The next significant discovery in this area was made by Alam et al. [100], when they illuminated ITO films at their plasma wavelength, and observed unimaginably high values of nonlinear coefficient $n_{2(eff)}$ and attenuation constant $\beta_{(eff)}$. The idea behind illumination at plasma wavelength λ_p was that being metal-like, ITO acquires ENZ (epsilon-near-zero) material behavior as the real part of its permittivity tends to zero at λ_p. They used a 310 nm-thick ITO film deposited on a glass substrate and illuminated it by a p-polarized light at various oblique angles of incidence and for various excitation wavelengths. Using the z-scan technique, they measured the effective nonlinear coefficient $n_{2(eff)} = \Delta n / I$ and the effective attenuation constant $\beta_{(eff)} = \Delta \alpha / I$. The authors found that both the parameters peaked at $\lambda_p = 1240$ nm and reduced at any wavelength longer or shorter than λ_p. They also found that $n_{2(eff)}$ and $\beta_{(eff)}$ gradually increased with the angle of incidence till $\theta = 60°$ and reduced sharply beyond that. The maximum value of $n_{2(eff)}$ achieved was 0.11 cm^2/GW at $\theta = 60°$ and $\lambda = 1240$ nm, dwarfing even the well-acknowledged As$_2$Se$_3$ chalcogenide glass ($\approx 10^{-5}$ cm^2/GW) [226]. In this way, it was observed that the effective nonlinear parameters depend on the excitation wavelength as well as the angle of incidence.

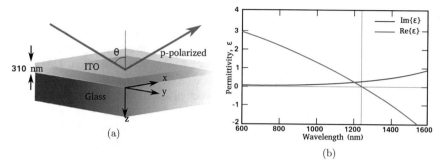

Fig. 4.25 a Schematic illustration of ITO thin film structure used by Alam et al., **b** permittivity of the film obtained by ellipsometry

Fig. 4.26 Transient response according to two-temperature model

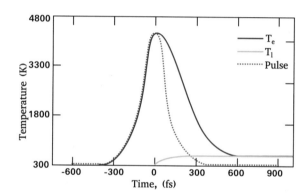

4.8.1 Hot Electron Phenomenon

The extraordinary nonlinear behavior of ITO is based on hot electron phenomenon and can be explained by the two-temperature model of laser heating. When an intense beam of ultrashort laser pulses hits a small particle or a small region of the sample, electrons absorb some part of the energy and rise to very high temperatures, much higher than the rest of the lattice, so these electrons are called *hot electrons*. Hot electrons are not in thermal equilibrium with the lattice. They stay in high-temperature state for a short duration and obey an altered Fermi–Dirac distribution [227]. The modified Fermi–Dirac distribution results in modified values of permittivity and susceptibilities. They lose the energy within 300–500 fs by electron–phonon interaction and the thermal equilibrium is achieved. Figure 4.26 shows how the electron temperature elevates with the growth of femtosecond laser pulse intensity. Even after the decay of the laser pulse, the electrons stay in for a short period of time and gradually lose energy to finally de-excite down to the lattice temperature and achieve thermal equilibrium. The red curve represents the input laser pulse, the blue one indicates the temperature variation of electrons with time, while the green one shows the same for the lattice. We take a pause to mention here that the graphs shown in Figs. 4.25b and

4.26 are schematics graphs and have not been plotted from data. Their purpose is to illustrate the variation in the shown quantities. To take a look at the actual graphs, the reader is advised to see Ref. [100]. In light of the above, ITO proves itself to be a potential candidate for on-chip nonlinear applications. However, ITO is not the only one of its kind. Aluminum zinc oxide (AZO) is another lucrative material of the same category and is a subject of investigation nowadays [228, 229]. Very recently, Carnemolla et al. [228, 229] have reported the spectral shift of twice the bandwidth in the refracted beam, on illumination an AZO thin film by its epsilon-near-zero wavelength [228]. In conclusion, TCOs hold the capability to revolutionize nonlinear photonics in the near future.

Chapter 5
Recent Advances in Zero-Index Photonics

In this chapter, we shall briefly describe a few important developments that have taken place in the zero-index photonics during the past decade. These developments are significant because they will show a path and become a foundation for zero-index technologies for the future.

5.1 Moitra's Design of Zero-Index Metamaterial

The first all-dielectric zero-index metamaterial proposed by Huang et al. was a 2D photonic crystal made up of a square array of dielectric rods [27]. A two-dimensional photonic crystal is infinite or practically very long in the third dimension compared to other two. Erecting very long rods vertically on a substrate is extremely challenging with nanoscale fabrication techniques. So Moitra et al. suggested a more plausible design [60] in which instead of erecting the rods vertically, they laid the rods horizontally on the substrate (see Fig. 5.1). In this way, one could have rods of any length without any difficulty, though there is a need for vertical spacer layers between the rods. In their design, the rods were made of silicon, the spacer layer was made of silica (SiO_2), and the whole device was embedded in polymethyl methacrylate (PMMA). Though Moitra's design is significant from the fabrication point of view, it lacks the compatibility to optical integration.

5.2 On-Chip Zero-Index Metamaterial

On-chip integration requires the structure to be periodic in the plane of the chip, and the TM mode operation requires the cylinders to be infinitely long. The first condition is achievable, but for the fulfillment of the second condition, Li et al. suggested an indirect yet effective way [92]. They confined the structure between two parallel metallic sheets, spread at the top and the bottom, as shown in Fig. 5.2a, so that the

N. Shankhwar and R. K. Sinha, *Zero Index Metamaterials*, https://doi.org/10.1007/978-981-16-0189-7_5

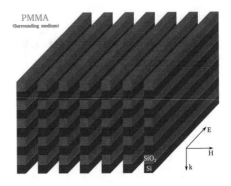

cylinders of a finite height can be made to behave as infinitely long rods and the "TM mode can be enforced." This is a very smart and effective design, which is perfectly suitable for on-chip fabrication. Figure 5.2b illustrates the steps involved in the fabrication of the ZIM. First, the silicon cylinders were etched on the silica (SiO$_2$) substrate. Then, a gold layer was deposited over the structure, occupying the top of the cylinder and the remaining surface of the silica substrate. After that, the cylinders were embedded in a layer of SU-8 matrix, followed by the final step involving the deposition of gold film over the SU-8 layer.

5.3   CMOS-Compatible Design of Zero-Index Metamaterial

The same group who developed the abovementioned on-chip zero-index metamaterial of silicon cylinders confined between metal sheets proposed an alternative design based on the complementary structure, i.e., air holes in a dielectric, for TE polarization (electric field oriented parallel to the substrate). The authors Vulis et al. labeled it "CMOS-compatible," as it can be made as a part of a standard 220 nm-thick silicon-on-insulator (SOI) device using the electron beam lithography technique, followed by reactive ion etching [114]. The design has been shown in Fig. 5.3, in which the SOI and the ZIM formed in it can be seen. This structure is very suitable for integrated photonics applications.

5.4   Zero-Index-Metamaterial-Based Nanoantenna

As explained in the previous chapters, the magnitude and phase of the electric field remain the same throughout the zero-index medium. Hence, if a point source exists inside the ZIM, all other points inside the ZIM and on its boundaries vibrate in coherence with the point source, making the ZIM act like an extended source of light. The longer boundary of the ZIM slab has more number of point sources than the shorter

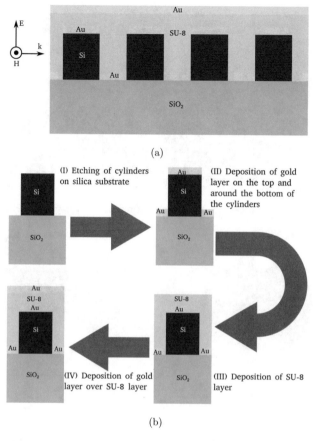

Fig. 5.2 On-chip design of zero-index metamaterial proposed by Li et al.

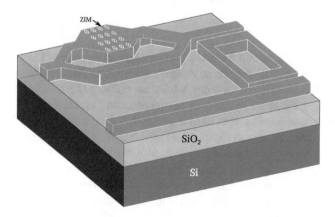

Fig. 5.3 Illustration of holes in dielectric-type ZIM as part of an SOI device, exhibiting the ZIM's CMOS compatibility

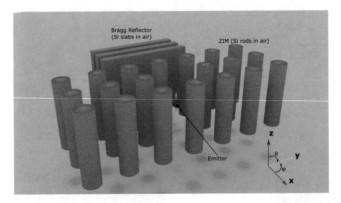

Fig. 5.4 Zero-index-metamaterial-based all-dielectric nanoantenna proposed by Shankhwar et al.

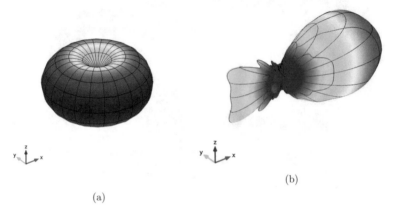

Fig. 5.5 **a** The original radiation pattern of the emitter. **b** The radiation pattern of the ZIM nanoantenna. The major radiation lobe is oriented toward +x-direction, indicating that most of the power is radiated in that direction

boundary, and consequently the former emits a greater amount of radiation than the latter. We exploited this property to design a zero-index-metamaterial-based highly directional all-dielectric nanoantenna, which can radiate in a particular direction, the optical power emitted by an otherwise omnidirectional quantum emitter [65]. The proposed nanoantenna has been shown in Fig. 5.4. It consists of a ZIM slab made up of a 7 × 3 array of silicon rods, backed by a Bragg reflector to prevent backward radiation. Figure 5.5 shows the radiation pattern of the nanoantenna, in which it can be observed that most of the radiation has been emitted in the +x-direction.

An important parameter called *directivity* gives the amount of power radiated by an antenna in a particular direction relative to the power radiated by a omnidirectional antenna. According to Krasnok et al. [230], Directivity (D) as function of spherical coordinates θ and ϕ is given by

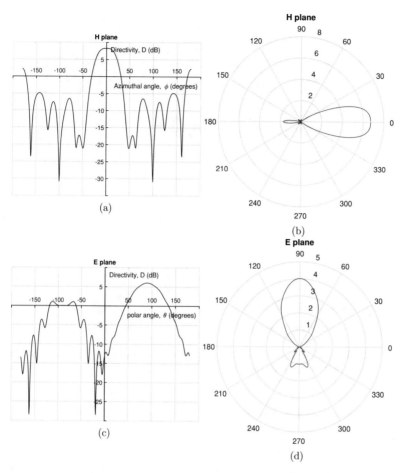

Fig. 5.6 Directivity D in H-plane—**a** Cartesian plot and **b** Polar plot. Directivity in E-plane—**c** Cartesian plot and **d** Polar plot

$$D(\theta, \phi) = \frac{4\pi p(\theta, \phi)}{P_{rad}} \tag{5.1}$$

where θ and ϕ, respectively, represent the polar and azimuthal coordinates of a spherical coordinate system, $p(\theta, \phi)$ is the power radiated in a particular direction, and P_{rad} is the total power radiated. We calculated directivity of the nanoantenna in both the E-plane and the H-plane.[1] The Cartesian and polar plots of directivity for both the planes have been shown in Fig. 5.6. The maximum directivity obtained in H-plane is 8.28 dB at $\phi = 0°$ and in E-plane is 6.04 dB at $\theta = 90°$.

---

[1] E-plane is the one containing the electric field vector while H-plane is the one containing the magnetic field vector. E-plane and H-plane are perpendicular to each other.

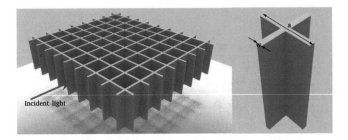

Fig. 5.7 Schematic illustration of the design of dielectric-vein-type zero-index metamaterial

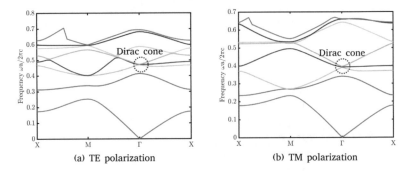

(a) TE polarization   (b) TM polarization

Fig. 5.8 The photonic band structures of dielectric-vein-type zero-index metamaterial for TM and TE polarizations, exhibiting Dirac cones in both the cases

5.5   Dielectric-Vein-Type Zero-Index Metamaterial

As discussed above, the dielectric-rod-based design was suitable for TM polarization, while the air holes in a dielectric-type structure were limited to TE polarization only. Hence, we felt the need for a design suitable for both polarizations and came up with a dielectric-vein-based structure, shown in Fig. 5.7. This structure can be tuned to operate at either of the two polarizations by slightly varying the width of the veins [64]. Figure 5.8 shows the photonic band structures for both TM and TE polarizations. It can be seen that a well-defined Dirac cone exists for both the modes, though at different frequencies. The Dirac point exists at $a/\lambda = 0.40$ for TM polarization, while at $a/\lambda = 0.46$ for TE polarization. The difference also lies in the vein width chosen in the two cases, which is $w = 0.22a$ for TM and $w = 0.3a$ for TE polarization.

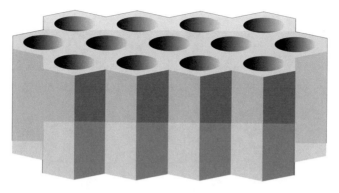

Fig. 5.9 Polarization-independent metamaterial proposed by Wang et al. [112]

5.6 Polarization-Independent Zero-Index Metamaterial

Wang et al. reported full polarization Dirac dispersion in *photonic hypercrystals* made by drilling air columns in an elliptic metamaterial. In other words, the structure reported by them is a polarization-independent zero-index metamaterial, exhibiting triple-degenerate Dirac point. The elliptic metamaterial is anisotropic in terms of permittivity $\epsilon_{host} = (\epsilon_{\parallel}, \epsilon_{\parallel}, \epsilon_{\perp})$, with $\epsilon_{\parallel} = 15.47$ and $\epsilon_{\perp} = 12.45$. The authors claim that such metamaterial can be constructed by layer-by-layer arrangement of natural materials [112]. The structure being a polarization-independent zero-index metamaterial, exhibited the property of directional radiation and the phenomenon of electromagnetic cloaking. This ZIM design is a significant landmark in the evolution of zero-index metamaterials (Fig. 5.9).

These were a few chief developments that have taken place in the arena of zero-index photonics and are worth mentioning. Besides these, there have been several lesser known yet important findings from time to time. We believe that, if substantially evolved, the zero-index photonics is capable of revolutionizing the photonic technology of the future.

Appendix
MPB Codes for Band Calculation of Various Photonic Crystals Described in This Book

A.1 Bulk silicon

bulkSi.ctl

```
; Compution of the bands at the X point for a quarter−wave stack Bragg
; mirror (this is the point that defines the band gap edges).

; the high and low indices:
(define−param n−lo 3.5)
(define−param n−hi 3.5)

; a quarter−wave stack
(define−param w−hi (/ n−lo (+ n−hi n−lo)))

; ld cell
(set! geometry−lattice (make lattice (size 1 no−size no−size)))
(set! default−material (make dielectric (index n−lo)))
(set! geometry
      (list
        (make cylinder
          (material (make dielectric (index n−hi)))
          (center 0 0 0) (axis 1 0 0)
          (radius infinity) (height w−hi))))

(set! k−points (list (vector3 −0.5 0 0)
                     (vector3 0.0 0 0)
                     (vector3 0.5 0 0)))
(set! k−points (interpolate 4 k−points))

(set−param! resolution 32)
(set−param! num−bands 3)

(run−tm output−hfield−y)
; Since TM and TE bands will be degenerate, so we calculate only TM
```

N. Shankhwar and R. K. Sinha, *Zero Index Metamaterials*,
https://doi.org/10.1007/978-981-16-0189-7

A.2 Bragg Mirror of Silicon-Air Quaterwave Stack

bragg.ctl

```
; Compute the bands at the X point for a quarter-wave stack Bragg
; mirror (this is the point that defines the band gap edges).

; the high and low indices:
(define-param n-lo 1.0)
(define-param n-hi 3.5)

(define-param w-hi (/ n-lo (+ n-hi n-lo))) ; a quarter-wave stack

; 1d cell
(set! geometry-lattice (make lattice (size 1 no-size no-size)))
(set! default-material (make dielectric (index n-lo)))
(set! geometry
      (list
        (make cylinder
          (material (make dielectric (index n-hi)))
          (center 0 0 0) (axis 1 0 0)
          (radius infinity) (height w-hi))))

(set! k-points (list (vector3 -0.5 0 0)
                                    (vector3 0.0 0 0)
                                    (vector3 0.5 0 0)))
(set! k-points (interpolate 4 k-points))

(set-param! resolution 32)
(set-param! num-bands 3)

(run-tm output-hfield-y)
; note that TM and TE bands are degenerate, so we only need TM
```

A.3 Square Lattice of Silicon Rods

sqrod.ctl

```
; PWM
; Band computation for a square lattice of Silicon rods

(set! num-bands 10)
(set! k-points (list (vector3 0.5 0.5 0)       ; M
                     (vector3 0.5 0 0)         ; X
                     (vector3 0 0 0)           ; Gamma
                     (vector3 0.5 0.5 0)       ; M
                )
)
(set! k-points (interpolate 9 k-points))
```

```
(set! geometry (list
                 (make cylinder (center 0.0 0 0) (radius 0.20)
 (height infinity) (material (make dielectric (epsilon 12.25))))
                 )
)
(set! geometry-lattice (make lattice (size 1 1 no-size)))
(set! resolution 32)
(run-te)
(run-tm (output-at-kpoint (vector3 0 0 0)
fix-efield-phase output-efield-z))
```

A.4 Square Lattice of Air Holes in Silicon Slab

sqholes.ctl

```
; PWM
; Band computation for a square lattice of air holes in silicon slab

(set! num-bands 17)
(set! k-points (list (vector3 0.5 0.5 0)        ; M
                     (vector3 0.5 0 0)          ; X
                     (vector3 0 0 0)            ; Gamma
                     (vector3 0.5 0.5 0)        ; M


                )
)
(set! k-points (interpolate 14 k-points))

(set! geometry (list
                 (make block
                     (center 0 0 0) (size 1 1 infinity)
                     (material (make dielectric (epsilon 12.25))))
                 (make cylinder
                     (center 0 0 0) (radius 0.315) (height infinity)
                     (material air))))
(set! geometry-lattice (make lattice (size 1 1 no-size)))
(set! resolution 32)
(run-tm)
(run-te)
```

A.5 Diamond Lattice of Silicon Spheres

diamond.ctl

```
; Diamond Lattice
(set! geometry-lattice (make lattice
                        (basis-size (sqrt 0.5) (sqrt 0.5) (sqrt 0.5))
                        (basis1 0 1 1)
                        (basis2 1 0 1)
                        (basis3 1 1 0)))
; Corners of irreducible Brilluoine Zone for Fcc lattice
; in cannonical order
(set! k-points (interpolate 4 (list
                        (vector3 0 0.5 0.5)         ; X
                        (vector3 0 0.625 0.375)     ; U
                        (vector3 0 0.5 0)           ; L
                        (vector3 0 0 0)             ; Gamma
                        (vector3 0 0.5 0.5)         ; X
                        (vector3 0.25 0.75 0.5)     ; W
                        (vector3 0.375 0.75 0.375)  ; K
                                                    )))
; define a couple of parameters (which we can set  from the command-line)
(define-param eps 12.25)      ; dielectric constant of sphere
(define-param r 0.2)          ; radius of sphere

(define diel (make dielectric (epsilon eps)))

; a diamong lattice has two atoms per unit cell:
(set! geometry (list (make sphere (center 0.125 0.125 0.125) (radius r)
           (material diel)) (make sphere (center -0.125 -0.125 -0.125)
           (radius r) (material diel))))

; A simple fcc lattice has only one sphere/object at the origin

(set-param! resolution 16)    ; use a 16x16x16 grid
(set-param! mesh-size 5)
(set-param! num-bands 5)

; run calculatio outputting electric field energy density at U point:
(run  (output-at-kpoint (vector3 0 0.625 0.375) output-dpwr))
```

NOTE: *The codes have been adapted from MPB tutorials available at the following links:*

http://ab-initio.mit.edu/mpb/doc/mpb.pdf

https://github.com/NanoComp/mpb/blob/master/examples/bragg.ctl

References

1. F.W. Sears, M.W. Zemansky, H.D. Young (University physics, Addison-Wesley, 1987)
2. R.P. Feynman, R.B. Leighton, M. Sands, *The Feynman Lectures on Physics, Vol. II: The New Millennium Edition: Mainly Electromagnetism and Matter*, vol. 2 (Basic books, 2011)
3. D.J. Griffiths, *Introduction to Electrodynamics* (AAPT, 2005)
4. A.W. Poyser, *Magnetism and Electricity: A Manual for Students in Advanced Classes* (Longmans, Green and Company, 1918)
5. M. Faraday, On some new electro-magnetical motions, and on the theory of magnetism. Quart. J. Sci. **12**, 74–96 (1821)
6. J.C. Maxwell, *A Treatise on Electricity and Magnetism*, vol. 1 (Clarendon Press, Oxford, 1873)
7. M.N.O. Sadiku, *Elements of Electromagnetics* (Oxford University Press, 2014)
8. M.R. Spiegel, *Schaum's Outline of Theory and Problems of Vector Analysis and An Introduction to Tensor Analysis* (McGraw-Hill, 1959)
9. R. Garg, *Analytical and Computational Methods in Electromagnetics* (Artech house, 2008)
10. A.K. Ghatak, *Optics* (Tata McGraw Hill, 2003)
11. A.K. Ghatak, K. Thyagarajan, *Optical Electronics* (Cambridge University Press, 1989)
12. R.P. Feynman, R.B. Leighton, M. Sands, *The Feynman Lectures on Physics, Vol. I: The New Millennium Edition: Mainly Mechanics, Radiation, and Heat*, vol. 1 (Basic books, 2011)
13. M. Born, E. Wolf, *Fundamentals of Optics* (M Nauka (Science), 1982)
14. F.A. Jenkins, H.E. White, *Fundamentals of Optics* (Tata McGraw-Hill Education, 1937)
15. S.A. Maier, *Plasmonics: Fundamentals and Applications* (Springer Science & Business Media, 2007)
16. V.G. Veselago, The electrodynamics of substances with simultaneously negative values of ϵ and μ. Sov. Phys. Uspekhi **10**(4), 509 (1968)
17. D.R. Smith, W.J. Padilla, D.C. Vier, S.C. Nemat-Nasser, S. Schultz, Composite medium with simultaneously negative permeability and permittivity. Phys. Rev. Lett. **84**(18), 4184 (2000)
18. D.R. Smith, J.B. Pendry, M.C.K. Wiltshire, Metamaterials and negative refractive index. Science **305**(5685), 788–792 (2004)
19. D.R. Smith, D.C. Vier, T. Koschny, C.M. Soukoulis, Electromagnetic parameter retrieval from inhomogeneous metamaterials. Phys. Rev. E **71**(3), 036617 (2005)
20. J.B. Pendry, A.J. Holden, W.J. Stewart, I. Youngs, Extremely low frequency plasmons in metallic mesostructures. Phys. Rev. Lett. **76**(25), 4773 (1996)
21. J.B. Pendry, A.J. Holden, D.J. Robbins, W.J. Stewart, Low frequency plasmons in thin-wire structures. J. Phys.: Condens. Matter **10**(22), 4785 (1998)
22. J.B. Pendry, A.J._ Holden, D.J. Robbins, W.J. Stewart, Magnetism from conductors and enhanced nonlinear phenomena. IEEE Trans. Microw Theory Tech. **47**(11), 2075–2084 (1999)
23. W. Cai, V.M. Shalaev. *Optical Metamaterials*, vol. 10 (Springer, 2010)

© The Editor(s) (if applicable) and The Author(s), under exclusive license to Springer Nature Singapore Pte Ltd. 2021
N. Shankhwar and R. K. Sinha, *Zero Index Metamaterials*,
https://doi.org/10.1007/978-981-16-0189-7

24. N. Engheta, R.W. Ziolkowski *Metamaterials: Physics and Engineering Explorations* (Wiley, 2006)
25. S. Enoch, G. Tayeb, P. Sabouroux, N. Guérin, P. Vincent, A metamaterial for directive emission. Phys. Rev. Lett. **89**(21), 213902 (2002)
26. C.L. Holloway, E.F. Kuester, J.A. Gordon, J.O. Hara, J. Booth, D.R. Smith, An overview of the theory and applications of metasurfaces: The two-dimensional equivalents of metamaterials. Antenn. Propag. Mag. IEEE **54**(2), 10–35 (2012)
27. X. Huang, Y. Lai, Z.H. Hang, H. Zheng, C.T. Chan, Dirac cones induced by accidental degeneracy in photonic crystals and zero-refractive-index materials. Nat. Mater. **10**(8), 582 (2011)
28. S. Jahani, Z. Jacob, All-dielectric metamaterials. Nat. Nanotechnol. **11**(1), 23–36 (2016)
29. K. Kishor, M.N. Baitha, R.K. Sinha, B. Lahiri.,Tunable negative refractive index metamaterial from v-shaped srr structure: fabrication and characterization. JOSA B **31**(7), 1410–1414 (2014)
30. S.A. Ramakrishna, Physics of negative refractive index materials. Rep. Prog. Phys. **68**(2), 449 (2005)
31. C. Rockstuhl, C. Menzel, T. Paul, T. Pertsch, F. Lederer, Light propagation in a fishnet metamaterial. Phys. Rev. B **78**(15), 155102 (2008)
32. J. Valentine, S. Zhang, T. Zentgraf, X. Zhang, Development of bulk optical negative index fishnet metamaterials: achieving a low-loss and broadband response through coupling. Proc. IEEE **99**(10), 1682–1690 (2011)
33. T-J. Yen, W.J. Padilla, N. Fang, D.C. Vier, D.R. Smith, J.B. Pendry, D.N. Basov, X. Zhang, Terahertz magnetic response from artificial materials. Science **303**(5663), 1494–1496 (2004)
34. H-K. Yuan, U.K. Chettiar, W. Cai, A.V. Kildishev, A. Boltasseva, V.P. Drachev, V.M. Shalaev, A negative permeability material at red light. Opt. Express **15**(3), 1076–1083 (2007)
35. Q. Zhao, J. Zhou, F. Zhang, D. Lippens, Mie resonance-based dielectric metamaterials. Mater. Today **12**(12), 60–69 (2009)
36. C. Kittel, *Introduction to Solid State Physics*, vol. 8 (Wiley, New York, 2004)
37. J.D. Joannopoulos, S.G. Johnson, J.N. Winn, R.D. Meade, *Photonic Crystals: Molding the Flow of Light* (Princeton university press, 2011)
38. E. Yablonovitch, Inhibited spontaneous emission in solid-state physics and electronics. Phys. Rev. Lett. **58**(20), 2059 (1987)
39. S. John, Strong localization of photons in certain disordered dielectric superlattices. Phys. Rev. Lett. **58**(23), 2486 (1987)
40. K. Sakoda, *Optical Properties of Photonic Crystals*, vol. 80 (Springer Science & Business Media, 2004)
41. Y. Cao, Z. Hou, Y. Liu, Convergence problem of plane-wave expansion method for phononic crystals. Phys. Lett. A **327**(2–3), 247–253 (2004)
42. S. Shi, C. Chen, D.W. Prather, Plane-wave expansion method for calculating band structure of photonic crystal slabs with perfectly matched layers. JOSA A **21**(9), 1769–1775 (2004)
43. K.S. Yee, J.S. Chen, The finite-difference time-domain (fdtd) and the finite-volume time-domain (fvtd) methods in solving maxwell's equations. IEEE Trans. Antenn. Propag. **45**(3), 354–363 (1997)
44. A. Taflove, S.C. Hagness, *Computational Electrodynamics: The Finite-Difference Time-Domain Method* (Artech house, 2005)
45. J.-M. Jin, *The Finite Element Method in Electromagnetics* (Wiley, 2015)
46. N.O. Matthew, *Sadiku, Numerical Techniques in Electromagnetics with MATLAB* (CRC Press, 2018)
47. COMSOL Inc. Comsol multiphysics
48. S.G. Johnson, J.D. Joannopoulos, Block-iterative frequency-domain methods for maxwell's equations in a planewave basis. Opt. Express **8**(3), 173–190 (2001)
49. Y.A. Arbabi, M.B. Horie, A. Faraon, Dielectric metasurfaces for complete control of phase and polarization with subwavelength spatial resolution and high transmission. Nat. Nanotechnol. **10**, 937–943 (2015)

50. N. Bonod, Silicon photonics: Large-scale dielectric metasurfaces. Nat. Mater. **14**(7), 664–665 (2015)
51. W.H. Bragg, W.L. Bragg, The reflection of x-rays by crystals. Proc. R. Soc. Lond. Ser. A, Contain. Pap. Math. Phys. Char. **88**(605), 428–438 (1913)
52. J. Cheng, D. Ansari-Oghol-Beig, H. Mosallaei, Wave manipulation with designer dielectric metasurfaces. Opt. Lett. **39**(21), 6285–6288 (2014)
53. A.B. Evlyukhin, S.M. Novikov, U. Zywietz, R.L. Eriksen, C. Reinhardt, S.I. Bozhevolnyi, B.N. Chichkov, Demonstration of magnetic dipole resonances of dielectric nanospheres in the visible region. Nano Lett. **12**(7), 3749–3755 (2012)
54. A. García-Etxarri, R. Gómez-Medina, L.S. Froufe-Pérez, C. López, L. Chantada, F. Scheffold, J. Aizpurua, M. Nieto-Vesperinas, J.J. Sáenz, Strong magnetic response of submicron silicon particles in the infrared. Opt. Express **19**(6), 4815–4826 (2011)
55. P. Gutruf, C. Zou, W. Withayachumnankul, M. Bhaskaran, S. Sriram, C. Fumeaux, Mechanically tunable dielectric resonator metasurfaces at visible frequencies. ACS Nano **10**(1), 133–141 (2015)
56. A.V. Kildishev, A. Boltasseva, V.M. Shalaev, Planar photonics with metasurfaces. Science **339**(6125), 1232009 (2013)
57. A. Krasnok, S. Makarov, M. Petrov, R. Savelev, P. Belov, Y. Kivshar, Towards all-dielectric metamaterials and nanophotonics, in *Proceedings SPIE 9502, Metamaterials X*, 9502:950203 (2015)
58. L. Lewin, The electrical constants of a material loaded with spherical particles. J. Inst. Electr. Eng.-Part III: Radio Commun. Eng. **94**(27), 65–68 (1947)
59. D. Lin, P. Fan, E. Hasman, M.L. Brongersma, Dielectric gradient metasurface optical elements. Science **345**(6194), 298–302 (2014)
60. P. Moitra, Y. Yang, Z. Anderson, I.I. Kravchenko, D.P. Briggs, J. Valentine, Realization of an all-dielectric zero-index optical metamaterial. Nat. Photon. **7**(10), 791 (2013)
61. P. Moitra, B.A. Slovick, W. Li, I.I. Kravchencko, D.P. Briggs, S. Krishnamurthy, J. Valentine, Large-scale all-dielectric metamaterial perfect reflectors. ACS Photon. **2**(6), 692–698 (2015)
62. N. Shankhwar, R.K. Sinha, Y. Kalra, S. Makarov, A. Krasnok, P. Belov, High-quality laser cavity based on all-dielectric metasurfaces. Photon. Nanostruct.-Fundam. Appl. **24**, 18–23 (2017)
63. N. Shankhwar, Y. Kalra, R.K. Sinha, Litao3 based metamaterial perfect absorber for terahertz spectrum. Superlatt. Microstruct. **111**, 754–759 (2017)
64. N. Shankhwar, Y. Kalra, R.K. Sinha, All dielectric zero-index metamaterial for te/tm polarization. J. Opt. **20**(11), 115101 (2018)
65. N. Shankhwar, Y. Kalra, Q. Li, R.K. Sinha, Zero-index metamaterial based all-dielectric nanoantenna. AIP Adv. **9**(3), 035115 (2019)
66. B. Slovick, Z.G. Yu, M. Berding, S. Krishnamurthy, Perfect dielectric-metamaterial reflector. Phys. Rev. B **88**(16), 165116 (2013)
67. Y. Yang, I.I. Kravchenko, D.P. Briggs, J. Valentine, All-dielectric metasurface analogue of electromagnetically induced transparency. Nat. Commun. **5**, 5753 (2014)
68. Yu. Nanfang, F. Capasso, Flat optics with designer metasurfaces. Nat. Mater. **13**(2), 139–150 (2014)
69. Y.F. Yu, A.Y. Zhu, R. Paniagua-Domínguez, Y.H. Fu, B. Luk'yanchuk, A.I. Kuznetsov, High-transmission dielectric metasurface with 2π phase control at visible wavelengths. Laser Photon. Rev. **9**(4), 412–418 (2015)
70. C.F. Bohren, D.R. Huffman, *Absorption and Scattering of Light by Small Particles* (Wiley, 2008)
71. P. Lalanne, D. Lemercier-Lalanne, On the effective medium theory of subwavelength periodic structures. J. Mod. Opt. **43**(10), 2063–2085 (1996)
72. M. Th Koschny, E.N.E. Kafesaki, C.M. Soukoulis, Effective medium theory of left-handed materials. Phys. Rev. Lett. **93**(10), 107402 (2004)
73. T.C. Choy, *Effective Medium Theory: Principles and Applications*, vol. 165 (Oxford University Press, 2015)

74. S. Foteinopoulou, Photonic crystals as metamaterials. Phys. B: Conden. Matter **407**(20), 4056–4061 (2012)

75. T. Antonakakis, R.V. Craster, S. Guenneau, Asymptotics for metamaterials and photonic crystals. Proc. R. Soc. A: Math. Phys. Eng. Sci. **469**(2152), 20120533 (2013)

76. F. Dominec, C. Kadlec, H. Němec, P. Kužel, F. Kadlec, Transition between metamaterial and photonic-crystal behavior in arrays of dielectric rods. Opt. express **22**(25), 30492–30503 (2014)

77. M.V. Rybin, D.S. Filonov, K.B. Samusev, P.A. Belov, Y.S. Kivshar, M.F. Limonov, Phase diagram for the transition from photonic crystals to dielectric metamaterials. Nat. Commun. **6**(1), 1–6 (2015)

78. E.E. Maslova, M.F. Limonov, M.V. Rybin, Transition between a photonic crystal and a meta-material with electric response in dielectric structures. JETP Lett. **109**(5), 340–344 (2019)

79. A. Berrier, M. Mulot, M. Swillo, M. Qiu, L. Thylén, A. Talneau, S. Anand, Negative refraction at infrared wavelengths in a two-dimensional photonic crystal. Phys. Rev. Lett. **93**(7), 073902 (2004)

80. E. Cubukcu, K. Aydin, E. Ozbay, S. Foteinopoulou, C.M. Soukoulis, Negative refraction by photonic crystals. Nature **423**(6940), 604–605 (2003)

81. C. Luo, S.G. Johnson, J.D. Joannopoulos, J.B. Pendry, All-angle negative refraction without negative effective index. Phys. Rev. B **65**(20), 201104 (2002)

82. C. Luo, S.G. Johnson, J.D. Joannopoulos, All-angle negative refraction in a three-dimensionally periodic photonic crystal. Appl. Phys. Lett. **81**(13), 2352–2354 (2002)

83. M. Notomi, Negative refraction in photonic crystals. Opt. Quantum Electron. **34**(1–3), 133–143 (2002)

84. P.V. Parimi, W.T. Lu, P. Vodo, J. Sokoloff, J.S. Derov, S. Sridhar, Negative refraction and left-handed electromagnetism in microwave photonic crystals. Phys. Rev. Lett. **92**(12), 127401 (2004)

85. L.-G. Wang, Z.-G. Wang, J.-X. Zhang, S.-Y. Zhu, Realization of dirac point with double cones in optics. Opt. Lett. **34**(10), 1510–1512 (2009)

86. D.R. Smith, J.B. Pendry, Homogenization of metamaterials by field averaging. J. Opt. Soc. Am. B **23**(3), 391–402 (2006)

87. Z. Szabo, G.-H. Park, R. Hedge, E.-P. Li, A unique extraction of metamaterial parameters based on kramers-kronig relationship. IEEE Trans. Microw. Theory Tech. **58**(10), 2646–2653 (2010)

88. X. Chen, T.M. Grzegorczyk, B.-I. Wu, J. Pacheco Jr., J.A. Kong, Robust method to retrieve the constitutive effective parameters of metamaterials. Phys. Rev. E **70**(1), 016608 (2004)

89. R.A. Shelby, D.R. Smith, S. Schultz, Experimental verification of a negative index of refraction. Science **292**(5514), 77–79 (2001)

90. A. Alu, M.G. Silveirinha, A. Salandrino, N. Engheta, Epsilon-near-zero metamaterials and electromagnetic sources: tailoring the radiation phase pattern. Phys. Rev. B **75**(15), 155410 (2007)

91. I. Liberal, N. Engheta, Near-zero refractive index photonics. Nat. Photon. **11**(3), 149 (2017)

92. Y. Li, S. Kita, P. Muñoz, O. Reshef, D.I. Vulis, M. Yin, M. Lončar, E. Mazur, On-chip zero-index metamaterials. Nat. Photon. **9**(11), 738 (2015)

93. K.C. Huang, M.L. Povinelli, J.D. Joannopoulos, Negative effective permeability in polaritonic photonic crystals. Appl. Phys. Lett. **85**(4), 543–545 (2004)

94. D. Korobkin, Y. Urzhumov, G. Shvets, Enhanced near-field resolution in midinfrared using metamaterials. JOSA B **23**(3), 468–478 (2006)

95. T. Taubner, D. Korobkin, Y. Urzhumov, G. Shvets, R. Hillenbrand, Near-field microscopy through a sic superlens. Science **313**(5793), 1595 (2006)

96. J.A. Schuller, R. Zia, T. Taubner, M.L. Brongersma, Dielectric metamaterials based on electric and magnetic resonances of silicon carbide particles. Phys. Rev. Lett. **99**(10), 107401 (2007)

97. N. Kinsey, C. DeVault, J. Kim, M. Ferrera, V.M. Shalaev, A. Boltasseva, Epsilon-near-zero al-doped zno for ultrafast switching at telecom wavelengths. Optica **2**(7), 616–622 (2015)

98. A. Anopchenko, S. Gurung, L. Tao, C. Arndt, H.W.H. Lee, Atomic layer deposition of ultra-thin and smooth al-doped zno for zero-index photonics. Mater. Res. Express **5**(1), 014012 (2018)
99. P. Kelly, L. Kuznetsova, Adaptive pre-shaping for ultrashort pulse control during propagation in azo/zno multilayered metamaterial at the epsilon-near-zero spectral point. OSA Contin. **3**(2), 143–152 (2020)
100. M.Z. Alam, I. De Leon, R.W. Boyd, Large optical nonlinearity of indium tin oxide in its epsilon-near-zero region. Science **352**(6287), 795–797 (2016)
101. J. Park, J.-H. Kang, X. Liu, M.L. Brongersma, Electrically tunable epsilon-near-zero (enz) metafilm absorbers. Sci. Rep. **5**, 15754 (2015)
102. M. Silveirinha, N. Engheta, Design of matched zero-index metamaterials using nonmagnetic inclusions in epsilon-near-zero media. Phys. Rev. B **75**(7), 075119 (2007)
103. A.M. Mahmoud, N. Engheta, Wave-matter interactions in epsilon-and-mu-near-zero structures. Nat. Commun. **5**(1), 1–7 (2014)
104. G. Dolling, C. Enkrich, M. Wegener, C.M. Soukoulis, S. Linden, Low-loss negative-index metamaterial at telecommunication wavelengths. Opt. Lett. **31**(12), 1800–1802 (2006)
105. E.D. Palik, *Handbook of Optical Constants of Solids*, vol. 3 (Academic, 1998)
106. M.I. Katsnelson, Zitterbewegung, chirality, and minimal conductivity in graphene. Eur. Phys. J. B-Condens. Matter Complex Syst. **51**(2), 157–160 (2006)
107. A.K. Geim, K.S. Novoselov, The rise of graphene. Nat. Mater. **6**(3), 183 (2007)
108. E. Kreyszig, *Advanced Engineering Mathematics*, 10th edn. (2009)
109. F. Zhang, G. Houzet, E. Lheurette, D. Lippens, M. Chaubet, X. Zhao, Negative-zero-positive metamaterial with omega-type metal inclusions. J. Appl. Phys. **103**(8), 084312 (2008)
110. P. Drude, Zur elektronentheorie der metalle. Annalen der physik **306**(3), 566–613 (1900)
111. C.T. Chan, Z.H. Hang, X. Huang, Dirac dispersion in two-dimensional photonic crystals, in *Advances in OptoElectronics* (2012)
112. J.-R. Wang, X.-D. Chen, F.-L. Zhao, J.-W. Dong, Full polarization conical dispersion and zero-refractive-index in two-dimensional photonic hypercrystals. Sci. Rep. **6**(1), 1–8 (2016)
113. L. Wang, H. Wang, S.K. Gupta, P. Wang, T. Lin, X. Liu, H. Lv, Second-harmonic phase-matching based on zero refractive index materials. Jpn. J. Appl. Phys. **58**(7), 072005 (2019)
114. D.I. Vulis, Y. Li, O. Reshef, P. Camayd-Muñoz, M. Yin, S. Kita, M. Lončar, E. Mazur, Monolithic cmos-compatible zero-index metamaterials. Opt. Express **25**(11), 12381–12399 (2017)
115. H. Tang, C. DeVault, P. Camayd-Munoz, D. Jia, Y. Liu, F. Du, O. Mello, Y. Li, E. Mazur, Low-loss zero-index materials (2020), arXiv:2004.01818
116. J.D. Ryder et al., *Networks, Lines and Fields* (1955)
117. J.A. Edminister, *Electric Circuits* (1965)
118. J.D. Kraus, R.J. Marhefka, A.S. Khan, *Antennas and Wave Propagation* (Tata McGraw-Hill Education, 2006)
119. C.A. Balanis, *Antenna Theory: Analysis and Design* (Wiley, 2016)
120. N. Engheta, Pursuing near-zero response. Science **340**(6130), 286–287 (2013)
121. O. Reshef, E. Giese, M.Z. Alam, I. De Leon, J. Upham, R.W. Boyd, Beyond the perturbative description of the nonlinear optical response of low-index materials. Opt. Lett. **42**(16), 3225–3228 (2017)
122. D. Halliday, R. Resnick, J. Walker, *Fundamentals of Physics* (Wiley, 2013)
123. H.J. Pain, R.T. Beyer, *The Physics of Vibrations and Waves* (Acoustical Society of America, 1993)
124. T. Miya, Y. Terunuma, T. Hosaka, T. Miyashita, Ultimate low-loss single-mode fibre at 1.55 μm. Electron. Lett. **15**(4), 106–108 (1979)
125. J.M. Senior, M.Y. Jamro, *Optical Fiber Communications: Principles and Practice* (Pearson Education, 2009)
126. G.P. Agrawal, *Fiber-Optic Communication Systems*, vol. 222 (Wiley, 2012)
127. International Telecommunication Union (ITU)

128. E.H. Turner, High-frequency electro-optic coefficients of lithium niobate. Appl. Phys. Lett. **8**(11), 303–304 (1966)

129. M. Luennemann, U. Hartwig, G. Panotopoulos, K. Buse, Electrooptic properties of lithium niobate crystals for extremely high external electric fields. Appl. Phys. B **76**(4), 403–406 (2003)

130. H. Okayama, Lithium niobate electro-optic switching, in *Optical Switching* (Springer, 2006), pp. 39–81

131. M. Silveirinha, N. Engheta, Tunneling of electromagnetic energy through subwavelength channels and bends using ε-near-zero materials. Phys. Rev. Lett. **97**(15), 157403 (2006)

132. J.S. Marcos, M.G. Silveirinha, N. Engheta, μ-near-zero supercoupling. Phys. Rev. B **91**(19), 195112 (2015)

133. V.C. Nguyen, L. Chen, K. Halterman, Total transmission and total reflection by zero index metamaterials with defects. Phys. Rev. Lett. **105**(23), 233908 (2010)

134. P. Alitalo, S. Tretyakov, Electromagnetic cloaking with metamaterials. Mater. Today **12**(3), 22–29 (2009)

135. D. Schurig, J.J. Mock, B.J. Justice, S.A. Cummer, J.B. Pendry, A.F. Starr, D.R. Smith, Metamaterial electromagnetic cloak at microwave frequencies. Science **314**(5801), 977–980 (2006)

136. J.B. Pendry, D. Schurig, D.R. Smith, Controlling electromagnetic fields. Science **312**(5781), 1780–1782 (2006)

137. E.P. Furlani, A. Baev, Electromagnetic analysis of cloaking metamaterial structures, in *Excerpt from the Proceedings of the COMSOL Conference* (2008)

138. N.B. Kundtz, D.R. Smith, J.B. Pendry, Electromagnetic design with transformation optics. Proc. IEEE **99**(10), 1622–1633 (2010)

139. X. Lin, H. Chen, Conformal transformation optics. Nat. Photon. **9**(1), 15–23 (2015)

140. W. Cai, U.K. Chettiar, A.V. Kildishev, V.M. Shalaev, Optical cloaking with metamaterials. Nat. Photon. **1**(4), 224–227 (2007)

141. S.A. Cummer, B.-I. Popa, D. Schurig, D.R. Smith, J. Pendry, Full-wave simulations of electromagnetic cloaking structures. Phys. Rev. E **74**(3), 036621 (2006)

142. J. Hao, W. Yan, M. Qiu, Super-reflection and cloaking based on zero index metamaterial. Appl. Phys. Lett. **96**(10), 101109 (2010)

143. W. Ying, J. Li, Total reflection and cloaking by zero index metamaterials loaded with rectangular dielectric defects. Appl. Phys. Lett. **102**(18), 183105 (2013)

144. H. Chu, Q. Li, B. Liu, J. Luo, S. Sun, Z.H. Hang, L. Zhou, Y. Lai, A hybrid invisibility cloak based on integration of transparent metasurfaces and zero-index materials. Light: Sci. Appl. **7**(1), 1–8 (2018)

145. Y. Huang, J. Li, Total reflection and cloaking by triangular defects embedded in zero index metamaterials. Adv. Appl. Math. Mech. **7**(2), 135–144 (2015)

146. S.S. Islam, M.R. Iqbal Faruque, M.T. Islam, A near zero refractive index metamaterial for electromagnetic invisibility cloaking operation. Materials **8**(8), 4790–4804 (2015)

147. J.A. Bossard, Y. Tang, D.H. Werner, T.S. Mayer, Genetic algorithm synthesis of planar zero index metamaterials for the infrared with application to electromagnetic cloaking, in *2007 IEEE Antennas and Propagation Society International Symposium* (IEEE, 2007), pp. 5555–5558

148. N. Shankhwar, Y. Kalra, R.K. Sinha, Dielectric veins type photonic crystal as a zero-index-metamaterial, in *Frontiers in Optics* (Optical Society of America, 2017), pp. FM3D–8

149. M. Navarro-Cía, M. Beruete, I. Campillo, M. Sorolla, Enhanced lens by ε and μ near-zero metamaterial boosted by extraordinary optical transmission. Phys. Rev. B **83**(11), 115112 (2011)

150. M. Navarro-Cía, M. Beruete, M. Sorolla, N. Engheta, Lensing system and fourier transformation using epsilon-near-zero metamaterials. Phys. Rev. B **86**(16), 165130 (2012)

151. V. Torres, B. Orazbayev, V. Pacheco-Peña, J. Teniente, M. Beruete, M. Navarro-Cía, M.S. Ayza, N. Engheta, Experimental demonstration of a millimeter-wave metallic enz lens based on the energy squeezing principle. IEEE Trans. Antenn. Propag. **63**(1), 231–239 (2014)

152. R.W. Ziolkowski, Propagation in and scattering from a matched metamaterial having a zero index of refraction. Phys. Rev. E **70**(4), 046608 (2004)
153. J.C. Soric, A. Alù, Longitudinally independent matching and arbitrary wave patterning using ε-near-zero channels. IEEE Trans. Microw. Theory Tech. **63**(11), 3558–3567 (2015)
154. S.M. Macneille, Beam splitter, July 9 1946. US Patent 2,403,731
155. M. Bayindir, B Temelkuran, and EJAPL Ozbay. Photonic-crystal-based beam splitters. Appl. Phys. Lett. **77**(24), 3902–3904 (2000)
156. C.-C. Chen, H.-D. Chien, P.-G. Luan, Photonic crystal beam splitters. Appl. Opt. **43**(33), 6187–6190 (2004)
157. X. Hongnan, D. Dai, Y. Shi, Ultra-broadband and ultra-compact on-chip silicon polarization beam splitter by using hetero-anisotropic metamaterials. Laser Photon. Rev. **13**(4), 1800349 (2019)
158. R. Halir, P. Cheben, J.M. Luque-González, J.D. Sarmiento-Merenguel, J.H. Schmid, G. Wangüemert-Pérez, D.-X. Xu, S. Wang, A. Ortega-Moñux, Í. Molina-Fernández, Ultra-broadband nanophotonic beamsplitter using an anisotropic sub-wavelength metamaterial. Laser Photo. Rev. **10**(6), 1039–1046 (2016)
159. O.T. Von Ramm, S.W. Smith, Beam steering with linear arrays. IEEE Trans. Biomed. Eng. **8**, 438–452 (1983)
160. D.F. Sievenpiper, J.H. Schaffner, H.J. Song, R.Y. Loo, G. Tangonan, Two-dimensional beam steering using an electrically tunable impedance surface. IEEE Trans. Antenn. Propag. **51**(10), 2713–2722 (2003)
161. J. Xu, J. Tang, Tunable prism based on piezoelectric metamaterial for acoustic beam steering. Appl. Phys. Lett. **110**(18), 181902 (2017)
162. D. Jia, Y. He, N. Ding, J. Zhou, D. Biao, W. Zhang, Beam-steering flat lens antenna based on multilayer gradient index metamaterials. IEEE Antenn. Wirel. Propag. Lett. **17**(8), 1510–1514 (2018)
163. V. Pacheco-Peña, V. Torres, B. Orazbayev, M. Beruete, M. Navarro-Cia, M. Sorolla, N. Engheta, Mechanical 144 ghz beam steering with all-metallic epsilon-near-zero lens antenna. Appl. Phys. Lett. **105**(24), 243503 (2014)
164. V. Pacheco-Peña, V. Torres, M. Beruete, M. Navarro-Cía, N. Engheta, ϵ-near-zero (enz) graded index quasi-optical devices: steering and splitting millimeter waves. J. Opt. **16**(9), 094009 (2014)
165. J.-P. Berenger et al., A perfectly matched layer for the absorption of electromagnetic waves. J. Comput. Phys. **114**(2), 185–200 (1994)
166. K. Artmann, Berechnung der seitenversetzung des totalreflektierten strahles. Annalen der Physik **437**(1–2), 87–102 (1948)
167. R.H. Renard, Total reflection: a new evaluation of the goos-hänchen shift. JOSA **54**(10), 1190–1197 (1964)
168. A. Ghatak, *Contemporary Optics* (Springer Science & Business Media, 2012)
169. Y. Xu, C. Ting Chan, H. Chen, Goos-hänchen effect in epsilon-near-zero metamaterials. Sci. Rep. **5**, 8681 (2015)
170. L. Boltzmann, Studien uber das gleichgewicht der lebenden kraft. Wissenschafiliche Abhandlungen **1**, 49–96 (1868)
171. J.W. Gibbs, *Elementary Principles in Statistical Mechanics: Developed with Especial Reference to the Rational Foundations of Thermodynamics* (C. Scribner's sons, 1902)
172. L.D. Landau, E. Lifshits, *Course of Theoretical Physics: Statistical Physics*
173. K. Thyagarajan, A. Ghatak, *Lasers: Fundamentals and Applications* (Springer Science & Business Media, 2010)
174. M. Planck, The theory of heat radiation, trans. *M. Masius, P. Blakiston's Son & Co, Philadelphia*, 1914
175. A. Beiser, *Concepts of Modern Physics* (Tata McGraw-Hill Education, 2003)
176. A.E. Siegman, Lasers university science books. Mill Valley CA **37**(208), 169 (1986)
177. R.G. Gould et al., The laser, light amplification by stimulated emission of radiation, in *The Ann Arbor Conference on Optical Pumping, the University of Michigan*, vol. 15 (1959), p. 92

178. T.H. Maiman, Stimulated optical radiation in ruby. Nature **187**(4736), 493–494 (1960)
179. C.H. Townes, *How the Laser Happened: Adventures of a Scientist* (Oxford University Press, 2002)
180. A. Javan, W.R. Bennett Jr., D.R. Herriott, Population inversion and continuous optical maser oscillation in a gas discharge containing a he-ne mixture. Phys. Rev. Lett. **6**(3), 106 (1961)
181. T. Kobayashi, T. Segawa, Y. Morimoto, T. Sueta, Novel-type lasers, emitting devices, and functional optical devices by controlling spontaneous emission, in *46th Fall Meeting of the Japanese Applied Physics Society* (1982)
182. J. McKeever, A. Boca, A.D. Boozer, J.R. Buck, H. Jeff Kimble, Experimental realization of a one-atom laser in the regime of strong coupling. Nature **425**(6955), 268–271 (2003)
183. R.W. Boyd, *Nonlinear Optics* (Academic, 2008)
184. eg P.A. Franken, A.E. Hill, C.W. el Peters, G. Weinreich. Generation of optical harmonics. Phys. Rev. Lett. **7**(4), 118 (1961)
185. J.E. Geusic, H.M. Marcos, L. Van Uitert, Laser oscillations in nd-doped yttrium aluminum, yttrium gallium and gadolinium garnets. Appl. Phys. Lett. **4**(10), 182–184 (1964)
186. W. Koechner, *Solid-State Laser Engineering*, vol. 1 (Springer, 2013)
187. A. Yariv, *Quantum Electronics* (Wiley, 1967)
188. P. Meystre, M. Sargent, *Three and Four Wave Mixing. In: Elements of Quantum Optics* (Springer, 1991)
189. K.O. Hill, D.C. Johnson, B.S. Kawasaki, R.I. MacDonald, cw three-wave mixing in single-mode optical fibers. J. Appl. Phys. **49**(10), 5098–5106 (1978)
190. S.E. Harris, M.K. Oshman, R.L. Byer, Observation of tunable optical parametric fluorescence. Phys. Rev. Lett. **18**(18), 732 (1967)
191. R.L. Byer, S.E. Harris, Power and bandwidth of spontaneous parametric emission. Phys. Rev. **168**(3), 1064 (1968)
192. Y. Chen, Solution to full coupled wave equations of nonlinear coupled systems. IEEE J. Quantum Electron. **25**(10), 2149–2153 (1989)
193. Y.H. Ja, Using the shooting method to solve boundary-value problems involving nonlinear coupled-wave equations. Opt. Quantum Electron. **15**(6), 529–538 (1983)
194. Y.H. Ja, Finite element method to solve the nonlinear coupled-wave equations for degenerate two-wave and four-wave mixing. Appl. Opt. **25**(23), 4306–4310 (1986)
195. A. Fiore, V. Berger, E. Rosencher, P. Bravetti, J. Nagle, Phase matching using an isotropic nonlinear optical material. Nature **391**(6666), 463–466 (1998)
196. S. Saltiel, Y.S. Kivshar, Phase matching in nonlinear χ (2) photonic crystals. Opt. Lett. **25**(16), 1204–1206 (2000)
197. D. Dimitropoulos, V. Raghunathan, R. Claps, B. Jalali, Phase-matching and nonlinear optical processes in silicon waveguides. Opt. express **12**(1), 149–160 (2004)
198. J.E. Midwinter, J. Warner, The effects of phase matching method and of uniaxial crystal symmetry on the polar distribution of second-order non-linear optical polarization. Br. J. Appl. Phys. **16**(8), 1135 (1965)
199. W.H. Besant, Chapter iii. The ellipse. *Conic Sections* (George Bell and Sons, London, 1907) p. 50,
200. R.W. Terhune, P.D. Maker, C.M. Savage, Optical harmonic generation in calcite. Phys. Rev. Lett. **8**(10), 404 (1962)
201. P.D. Maker, R.W. Terhune, M. Nisenoff, C.M. Savage, Effects of dispersion and focusing on the production of optical harmonics. Phys. Rev. Lett. **8**(1), 21 (1962)
202. R.L. Byer, Quasi-phasematched nonlinear interactions and devices. J. Nonlinear Opt. Phys. Mater. **6**(04), 549–592 (1997)
203. M.M. Fejer, G.A. Magel, D.H. Jundt, R.L. Byer, Quasi-phase-matched second harmonic generation: tuning and tolerances. IEEE J. Quantum Electron. **28**(11), 2631–2654 (1992)
204. G.A.R.O. Khanarian, R.A. Norwood, D. Haas, B. Feuer, D. Karim, Phase-matched second-harmonic generation in a polymer waveguide. Appl. Phys. Lett. **57**(10), 977–979 (1990)
205. M. Yamada, N. Nada, M. Saitoh, K. Watanabe, First-order quasi-phase matched linbo3 waveguide periodically poled by applying an external field for efficient blue second-harmonic generation. Appl. Phys. Lett. **62**(5), 435–436 (1993)

206. J.A. Armstrong, N. Bloembergen, J. Ducuing, P.S. Pershan, Interactions between light waves in a nonlinear dielectric. Phys. Rev. **127**(6), 1918 (1962)
207. M. Cazzanelli, J. Schilling, Second order optical nonlinearity in silicon by symmetry breaking. Appl. Phys. Rev. **3**(1), 011104 (2016)
208. P.D. Maker, R.W. Terhune, C.M. Savage, Intensity-dependent changes in the refractive index of liquids. Phys. Rev. Lett. **12**(18), 507 (1964)
209. E.L. Buckland, R.W. Boyd, Electrostrictive contribution to the intensity-dependent refractive index of optical fibers. Opt. Lett. **21**(15), 1117–1119 (1996)
210. D.W. Garvey, Q. Li, M.G. Kuzyk, C.W. Dirk, S. Martinez, Sagnac interferometric intensity-dependent refractive-index measurements of polymer optical fiber. Opt. Lett. **21**(2), 104–106 (1996)
211. O. Reshef, I. De Leon, M.Z. Alam, R.W. Boyd, Nonlinear optical effects in epsilon-near-zero media. Nat. Rev. Mater. **4**(8), 535–551 (2019)
212. N.K. Hon, R. Soref, B. Jalali, The third-order nonlinear optical coefficients of si, ge, and si1-x ge x in the midwave and longwave infrared. J. Appl. Phys. **110**(1), 9 (2011)
213. P.L. Kelley, Self-focusing of optical beams. Phys. Rev. Lett. **15**(26), 1005 (1965)
214. A. Shabat, V. Zakharov, Exact theory of two-dimensional self-focusing and one-dimensional self-modulation of waves in nonlinear media. Sov. Phys. JETP **34**(1), 62 (1972)
215. Y. Shen, Self-focusing: experimental. Prog. Quantum Electron. **4**, 1–34 (1975)
216. S.A. Akhmanov, A.P. Sukhorukov, R.V. Khokhlov, Self-focusing and diffraction of light in a nonlinear medium. Sov. Phys. USPEKHI **10**(5), 609 (1968)
217. R.Y. Chiao, E. Garmire, C.H. Townes, Self-trapping of optical beams. Phys. Rev. Lett. **13**(15), 479 (1964)
218. A.S. Kewitsch, A. Yariv, Self-focusing and self-trapping of optical beams upon photopolymerization. Opt. Lett. **21**(1), 24–26 (1996)
219. M. Mitchell, M. Segev, Self-trapping of incoherent white light. Nature **387**(6636), 880–883 (1997)
220. J.L. Humphrey, D. Kuciauskas, Optical susceptibilities of supported indium tin oxide thin films. J. Appl. Phys. **100**(11), 113123 (2006)
221. K.L. Chopra, S. Major, D.K. Pandya, Transparent conductor - a status review. Thin solid films **102**(1), 1–46 (1983)
222. I. Hamberg, C.G. Granqvist, Evaporated sn-doped in2o3 films: Basic optical properties and applications to energy-efficient windows. J. Appl. Phys. **60**(11), R123–R160 (1986)
223. R.B.H. Tahar, T. Ban, Y. Ohya, Y. Takahashi, Tin doped indium oxide thin films: Electrical properties. J. Appl. Phys. **83**(5), 2631–2645 (1998)
224. K. Zhang, F. Zhu, C.H.A. Huan, A.T.S. Wee, Indium tin oxide films prepared by radio frequency magnetron sputtering method at a low processing temperature. Thin Solid Films **376**(1–2), 255–263 (2000)
225. H.I. Elim, W. Ji, F. Zhu, Carrier concentration dependence of optical kerr nonlinearity in indium tin oxide films. Appl. Phys. B **82**(3), 439–442 (2006)
226. J. Troles, F.M.P.V.H.G. Smektala, M. Guignard, P. Houizot, V. Nazabal, H. Zeghlache, G. Boudebs, V. Couderc, Third and second order non linear optical properties of chalcogenide glasses on bulk and fibers, in *Proceedings of 2005 7th International Conference Transparent Optical Networks*, vol. 2 (IEEE, 2005), pp. 242–244
227. P.J. Price, Calculation of hot electron phenomena. Solid-State Electron. **21**(1), 9–16 (1978)
228. E.G. Carnemolla, V. Bruno, L. Caspani, M. Clerici, S. Vezzoli, T. Roger, C. DeVault, J. Kim, A. Shaltout, V. Shalaev et al., Giant nonlinear frequency shift in epsilon-zero aluminum zinc oxide thin films, in *CLEO: Science and Innovations* (Optical Society of America, 2018), pp. SM4D–7
229. E.G. Carnemolla, L. Caspani, C. DeVault, M. Clerici, S. Vezzoli, V. Bruno, V.M. Shalaev, D. Faccio, A. Boltasseva, M. Ferrera, Degenerate optical nonlinear enhancement in epsilon-near-zero transparent conducting oxides. Opt. Mater. Express **8**(11), 3392–3400 (2018)
230. A.E. Krasnok, I.S. Maksymov, A.I. Denisyuk, P.Al. Belov, A.E. Miroshnichenko, C.R. Simovski, Y.S. Kivshar, Optical nanoantennas. Phys.-Uspekhi **56**(6), 539 (2013)

Printed in the United States
by Baker & Taylor Publisher Services